Beyond the Stars: Quantum Mechanics in Space

N.B. Singh

DEDICATION

To Nature,

I dedicate this book to you, the source of all life. You are my inspiration, my teacher, and my friend.

Thank you for teaching me about the beauty of the world around me. Thank you for showing me the power of the natural world. Thank you for giving me a sense of peace and tranquillity.

I promise to do my part to protect you and your many wonders. I will teach my children about the importance of conservation and sustainability. I will work to make the world a better place for all living things.

Thank you for everything, Nature.

With love,

N.B Singh

Contents

IV Quantum Technology for Space Exploration 71

PREFACE

In the ever-expanding tapestry of the cosmos, the marriage of quantum mechanics and space science unfolds a narrative that transcends the boundaries of our understanding. "Beyond the Stars: Quantum Mechanics in Space" is an exploration into the profound connections between the microscopic world of quantum particles and the vast reaches of the universe.

As we delve into the pages of this book, we embark on a journey that traverses the intricacies of quantum entanglement, the mysteries of black holes, and the cosmic dance of particles and waves. The chapters within are a testament to the symbiotic relationship between the quantum realm and the celestial bodies that grace our night skies.

From the quantum superposition that defines the evolution of galactic structures to the enigmatic dance of particles near black holes, each concept is a brushstroke painting a portrait of the universe in the language of mathematics and physics. We explore the quantum technologies that enhance our space exploration endeavors and unravel the quantum aspects of phenomena ranging from active galactic nuclei to the cosmic evolution itself.

This book is an invitation to readers, whether seasoned physicists or curious minds, to embark on a journey that transcends the ordinary boundaries of space and time. It is a testament to the beauty of the universe and the power of human intellect in deciphering its secrets.

In the chapters that follow, we encounter the quantum fingerprints in the formation of galaxies, witness the entanglement of particles across cosmic distances, and ponder the implications of quantum mechanics in the vastness of space. The text is adorned with mathematical formulations, illustrations, and examples to guide the reader through the intricate dance of quantum phenomena in the cosmos.

Let "Beyond the Stars: Quantum Mechanics in Space" be your companion in the exploration of the extraordinary, where the language of quantum mechanics becomes the key to unlocking the mysteries that lie far beyond the stars.

Introduction

Welcome to the cosmic odyssey of "Beyond the Stars: Quantum Mechanics in Space." In this journey, we'll explore the fascinating intersection of quantum mechanics and the celestial wonders that grace our universe.

Unveiling the Quantum Cosmos

At the heart of the cosmos lies the enigma of quantum mechanics. From the peculiarities of particle behavior to the entangled dance of particles across light-years, we are about to unravel the cosmic mysteries governed by the laws of quantum physics.

Quantum Superposition in Celestial Bodies

Imagine a star simultaneously existing in multiple states, radiating quantum possibilities across the cosmic expanse. This is the quantum superposition, a phenomenon that extends its influence to the very fabric of space.

Quantum Orbits and Celestial Ballet

Our celestial journey takes us to the graceful ballet of planets, where quantum orbits dictate the cosmic dance. The elegance of planetary motion is a testament to the harmonious interplay between gravity and quantum principles.

Quantum Tunneling in Stellar Cores

Deep within the cores of stars, quantum tunneling orchestrates the fusion processes that light up our night sky. Witness the breathtaking spectacle of quantum particles overcoming energy barriers, enabling the celestial furnaces to shine.

Black Holes: Quantum Abyss of Infinite Curvature

Venturing into the depths of space, we encounter black holes, celestial behemoths where quantum mechanics meets the infinite curvature of spacetime. Explore the quantum mysteries that shroud these cosmic enigmas.

Hawking Radiation: Quantum Whispers from Event Horizons

At the event horizons of black holes, quantum whispers manifest as Hawking radiation. Discover how quantum fluctuations near the cosmic abyss lead to the gradual evaporation of these celestial giants.

Quantum Communication Across Cosmic Distances

In the vastness of space, communication becomes a quantum challenge. Delve into the potential of quantum entanglement to revolutionize interstellar communication, connecting celestial outposts across unimaginable distances.

Quantum Entanglement and Extraterrestrial Networks

Imagine an extraterrestrial network where quantum-entangled particles enable instantaneous communication between distant star systems. The quantum realm becomes the backbone of celestial information highways.

Quantum Sensors on Cosmic Voyages

Embark on cosmic voyages equipped with advanced quantum sensors. These technologies promise to enhance our understanding of the universe by detecting subtle quantum phenomena, from gravitational waves to cosmic microwave background radiation.

Quantum Interference in Stellar Interferometry

In the realm of stellar interferometry, quantum interference plays a pivotal role in extracting intricate details about distant stars. Witness the quantum dance of photons as they combine and interfere to reveal cosmic secrets.

Conclusion: Navigating Quantum Horizons

As we conclude this introduction, we stand on the brink of quantum horizons beyond the stars. The journey ahead invites us to navigate the quantum fabric of space, unveiling the harmonies that resonate across the cosmic symphony.

Quantum Harmony in Celestial Symphony

In the celestial symphony, quantum harmony echoes through the cosmic tapestry. Our quest is to explore, comprehend, and marvel at the interwoven threads of quantum mechanics that shape the grand spectacle "Beyond the Stars."

Part I

Foundations of Quantum Mechanics

Chapter 1

Quantum Revolution

1.1 The Quantum Revolution

Welcome to a chapter that transcends celestial boundaries – "The Quantum Revolution." In this section, we delve into the transformative principles of quantum mechanics, reshaping our understanding of the cosmos.

1.1.1 Wave-Particle Duality in Celestial Bodies

At the heart of the Quantum Revolution lies the profound concept of wave-particle duality. Consider a photon emitted by a distant star. Quantum mechanics reveals that this seemingly simple particle behaves both as a discrete entity and a wave, creating ripples across the cosmic fabric.

The de Broglie wavelength equation captures this duality:

$$\lambda = \frac{h}{p}$$

where λ is the wavelength, h is the Planck constant, and p is the momentum.

1.1.2 Quantum Superposition in Stellar Phenomena

Stars, those cosmic luminaries, exist in states of quantum superposition. The very nature of these celestial entities challenges our classical intuitions. Mathematically, the superposition principle is expressed as:

$$\Psi = c_1\psi_1 + c_2\psi_2$$

where Ψ is the overall wavefunction, c_1 and c_2 are complex coefficients, and ψ_1 and ψ_2 are individual wavefunctions.

1.1.3   Entanglement Across Galactic Distances

Entanglement, a quantum phenomenon, extends its influence across vast cosmic distances. Imagine two entangled particles, separated by galactic expanses, instantaneously mirroring each other's states. This celestial entanglement challenges classical notions.

The entangled state is represented as:

$$|\Psi\rangle = \frac{|00\rangle + |11\rangle}{\sqrt{2}}$$

where $|00\rangle$ and $|11\rangle$ denote the possible states of the entangled particles.

1.1.4   Quantum Information and Cosmic Communication

As we traverse the cosmos, the concept of quantum information becomes pivotal. Quantum bits, or qubits, emerge as carriers of celestial messages. Quantum communication protocols promise secure transmissions across cosmic distances.

Quantum key distribution, a cornerstone of celestial communication, safeguards information from cosmic eavesdroppers.

1.1.5   Quantum Computing in Celestial Exploration

Our cosmic journey leads us to the frontier of celestial exploration with quantum computing. Celestial simulations, navigational algorithms, and complex computations find new frontiers with quantum devices.

Shor's Algorithm, a quantum marvel, stands ready to decrypt celestial secrets hidden in classical cryptography. This computational power becomes our celestial key.

1.1.6 Conclusion: Quantum Horizons Unveiled

"The Quantum Revolution" opens celestial horizons veiled in classical mysteries. This section serves as a cosmic preamble, inviting exploration into the quantum cosmos "Beyond the Stars."

In the symphony of the cosmos, quantum harmonies resonate, creating celestial melodies that echo through the vastness of space. As we gaze beyond the stars, the Quantum Revolution beckons us to explore the harmonic threads that weave through the cosmic tapestry.

1.2 Quantum Principles

The quantum revolution marked a profound shift in our understanding of the fundamental principles governing the behavior of particles at the quantum level. In this section, we delve into the key concepts that form the bedrock of quantum mechanics.

1.2.1 Superposition Principle

One of the cornerstones of quantum mechanics is the principle of superposition. Mathematically, it is expressed as follows:

$$|\psi\rangle = \alpha |0\rangle + \beta |1\rangle$$

Here, $|0\rangle$ and $|1\rangle$ represent the basis states, and α and β are complex coefficients. This formula illustrates how a quantum system can exist in multiple states simultaneously.

1.2.2 Entanglement

Entanglement is another intriguing quantum phenomenon. Consider two entangled particles A and B. The state of the combined system cannot be described independently for each particle; instead, it is given by:

$$|\Psi\rangle = \frac{1}{\sqrt{2}}(|0\rangle_A |1\rangle_B - |1\rangle_A |0\rangle_B)$$

The entangled particles become interconnected, and the state of one instantly influences the state of the other, regardless of the distance between them.

1.2.3   Quantum Operators

In quantum mechanics, operators play a central role in describing physical observables. The Hermitian operator corresponding to an observable is associated with a measurable quantity. For instance, the position of a particle is represented by the position operator X, and its momentum by the momentum operator P.

$$[X, P] = XP - PX = i\hbar$$

This commutation relation is fundamental to the uncertainty principle.

1.2.4   Uncertainty Principle

Heisenberg's uncertainty principle asserts that certain pairs of properties, such as position and momentum, cannot be simultaneously known precisely. Mathematically, it is expressed as:

$$\Delta x \cdot \Delta p \geq \frac{\hbar}{2}$$

This principle places a fundamental limit on the precision with which certain pairs of properties can be known.

1.2.5   Example: Quantum Superposition in a Qubit

Let's consider a quantum bit or qubit. The superposition principle allows the qubit to exist in a linear combination of the basis states $|0\rangle$ and $|1\rangle$:

$$|\psi\rangle = \alpha |0\rangle + \beta |1\rangle$$

where $|\alpha|^2 + |\beta|^2 = 1$. This flexibility forms the basis for quantum parallelism and quantum computation.

1.2.6 Example: Bell State Entanglement

A famous example of entanglement is the creation of Bell states. Consider two qubits A and B. The Bell state $|\Phi^+\rangle$ is given by:

$$|\Phi^+\rangle = \frac{1}{\sqrt{2}}(|0\rangle_A |0\rangle_B + |1\rangle_A |1\rangle_B)$$

Measuring one qubit instantly determines the state of the other, showcasing the non-local and interconnected nature of entanglement.

1.3 Quantum Entanglement

Quantum entanglement is a phenomenon that lies at the heart of quantum mechanics, defying classical intuition and challenging our understanding of the fundamental nature of reality. In this section, we embark on a journey into the mesmerizing world of entanglement, exploring its mathematical formulations, experimental observations, and the profound implications it carries for the fabric of spacetime.

1.3.1 Foundations of Entanglement

At the core of quantum entanglement is the concept of superposition, where quantum systems can exist in multiple states simultaneously. Entanglement takes this notion a step further by establishing correlations between entangled particles, such that the state of one particle instantaneously influences the state of another, regardless of the distance separating them.

1.3.2 Mathematics of Entanglement

Mathematically, entanglement is described using the concept of composite quantum states. Consider two entangled particles A and B. The joint state of the system is expressed as:

$$|\Psi\rangle = \frac{1}{\sqrt{2}}(|0\rangle_A |1\rangle_B - |1\rangle_A |0\rangle_B)$$

This Bell state illustrates how the particles become entwined, creating a singular, non-factorizable quantum state.

1.3.3 Entanglement and Measurement

The phenomenon of entanglement manifests most profoundly when measurements are performed on the entangled particles. For example, if we measure the spin of one particle along a certain axis and find it to be 'up,' the spin of the other particle, measured along the same axis, will instantaneously be 'down,' and vice versa. This instantaneous correlation defies classical notions of causality and communication.

1.3.4 Bell Inequality and Tests of Entanglement

The violation of Bell inequalities provides a powerful tool for experimentally confirming the existence of entanglement. Bell's theorem establishes limits on the correlations that can exist between classical particles, and violations of these limits serve as empirical evidence for the non-classical nature of entanglement.

1.3.5 Example: EPR Paradox

The EPR paradox, formulated by Einstein, Podolsky, and Rosen, laid the conceptual groundwork for understanding the counterintuitive nature of entanglement. In their famous paper, they argued that quantum mechanics is incomplete and proposed hidden variables to explain the correlations observed in entangled systems. However, subsequent experiments, such as those conducted by Alain Aspect, have consistently supported the predictions of quantum mechanics and refuted hidden variable theories.

1.3.6 Quantum Teleportation

Entanglement also finds practical applications in quantum teleportation, a process by which the exact state of one particle can be transmitted to another particle at a distant location. This process relies on the entangled correlations between particles, showcasing the potential for quantum technologies to revolutionize communication and computation.

Part II

Quantum Mechanics in Celestial Bodies

Chapter 2

Quantum Stars

2.1 Quantum Processes in Stellar Cores

As we journey beyond our terrestrial confines and explore the depths of celestial bodies, the application of quantum mechanics takes on a profound significance. In this section, we delve into the quantum processes that govern the behavior of matter in the cores of stars, unraveling the mysteries of nuclear reactions, energy generation, and the delicate balance that sustains these cosmic entities.

2.1.1 Nuclear Reactions in Extreme Environments

Stellar cores provide an arena for extreme conditions where nuclear reactions, governed by quantum principles, dictate the fate and evolution of stars. The fundamental process underlying stellar energy production is nuclear fusion, where light nuclei combine to form heavier elements, releasing vast amounts of energy in the process. This phenomenon is encapsulated by the famous equation $E = mc^2$, highlighting the conversion of mass into energy.

2.1.2   Quantum Tunneling and Fusion

The initiation of nuclear fusion in stellar cores relies on the quantum tunneling effect. Quantum particles overcome the electrostatic repulsion barrier through tunneling, allowing them to approach closely enough for the strong nuclear force to bind them together. The probability of this tunneling is described by the Gamow factor, providing a crucial quantum mechanical insight into the likelihood of nuclear fusion events.

2.1.3   Stellar Equilibrium and Degeneracy Pressure

Quantum degeneracy pressure plays a pivotal role in maintaining the equilibrium of stellar cores. As gravitational forces attempt to collapse the star, quantum effects prevent further compression by enforcing Pauli exclusion principles. This phenomenon, known as degeneracy pressure, counteracts gravitational forces and stabilizes stars, defining their size and density.

2.1.4   Energy Transport: Quantum Convection and Radiation

Within stellar cores, the transport of energy involves quantum convection and radiation. Quantum convection occurs as hot, less dense plasma rises, and cooler, denser plasma sinks, creating a dynamic energy transfer process. Additionally, radiation, modeled quantum mechanically as packets of energy called photons, contributes to the outward flow of energy from the stellar core.

2.1.5   Quantum Chaos in Supernovae

In the explosive finale of massive stars, quantum chaos takes center stage during supernova events. The collapse and subsequent explosion generate conditions where quantum effects become prominent. Neutrinos, elusive quantum particles, play a crucial role in carrying away energy and influencing the dynamics of these cataclysmic events.

2.1.6   Stellar Quantum States and Neutrino Emission

The final stages of stellar evolution involve intricate quantum states, where the collapse of stellar cores leads to the formation of neutron stars or black holes. Neutrino emission, a quantum process resulting from these collapses, provides valuable insights into the quantum properties of matter in extreme gravitational fields.

2.2   Quantum Entanglement in Binary Star Systems

As we turn our gaze to the celestial dance of binary star systems, the profound influence of quantum entanglement on the behavior of these stellar companions becomes evident. Quantum entanglement, a fundamental aspect of quantum mechanics, introduces a unique interconnectedness that transcends classical boundaries.

2.2.1   Understanding Quantum Entanglement

Quantum entanglement describes the phenomenon where two or more particles become interconnected in such a way that the state of one particle instantly influences the state of the other, regardless of the distance separating them. This entanglement defies classical intuition and forms the bedrock of quantum correlations.

2.2.2   Mathematics of Quantum Entanglement

To mathematically describe the entangled states of particles in a binary star system, we turn to the formalism of quantum mechanics. Let $|\psi\rangle$ represent the joint state of the binary stars, and the entanglement is characterized by a joint quantum state that cannot be factored into individual states: $|\psi\rangle \neq |\psi_1\rangle \otimes |\psi_2\rangle$.

2.2.3   Entangled States in Binary Systems

Consider a binary star system with two entangled particles. The entangled state can be expressed as a superposition of states, such as $|\psi\rangle = \alpha\,|00\rangle + \beta\,|11\rangle$, where α and β are probability amplitudes. The entangled state reflects correlations that classical physics cannot account for.

2.2.4   Quantum Measurement and Correlations

When a quantum measurement is performed on one particle in an entangled state, the outcome is instantaneously correlated with the state of the other, regardless of the spatial separation. This non-local correlation challenges classical notions of causality and provides a fascinating avenue for exploration in binary star systems.

2.2.5   Quantum Bell Inequality and Binary Stars

In the context of binary stars, researchers may employ Bell inequalities to test the boundaries of quantum entanglement. Violations of Bell inequalities imply non-classical correlations, offering a window into the intricacies of entangled states within binary star systems.

2.2.6   EPR Paradox Revisited

The EPR paradox, proposed by Einstein, Podolsky, and Rosen, questioned the completeness of quantum mechanics. In the realm of binary star systems, the revisitation of the EPR paradox becomes pertinent as entanglement reveals the inseparable nature of particles across vast cosmic distances.

2.2.7   Quantum Teleportation in Binary Star Communication

Exploring the potential applications of quantum entanglement in binary star systems, the concept of quantum teleportation emerges. Could quantum entanglement be harnessed for instantaneous communication between binary stars? The quantum communication possibilities in these celestial relationships open avenues for theoretical exploration.

2.2.8 Quantum Entanglement's Influence on Binary Star Dynamics

As binary stars orbit each other, quantum entanglement introduces a layer of complexity to their dynamics. The intertwined quantum states of the stars may impact their evolutionary paths, orbital parameters, and even the outcomes of stellar interactions.

2.2.9 Quantum Coherence in Binary Star Formation

In the genesis of binary star systems, the principles of quantum coherence may play a crucial role. Coherent quantum states could influence the formation process, leading to unique configurations and behaviors in binary star systems.

2.2.10 Quantum Correlations Across Cosmic Distances

The exploration of quantum entanglement in binary star systems extends our understanding of correlations across cosmic distances. The non-local nature of entanglement challenges conventional astrophysical models and invites a reexamination of the interconnectedness of celestial entities.

2.2.11 Quantum Information Transfer in Binary Stars

Within the entangled quantum states of binary stars lies the potential for information transfer beyond classical limits. Understanding how quantum information is encoded and decoded in these systems opens new avenues for research in quantum astrophysics.

2.2.12   Quantum Decoherence in Binary Star Interactions

While quantum entanglement introduces remarkable coherence, the interactions within binary star systems may also lead to decoherence. Investigating the factors influencing quantum decoherence becomes crucial for comprehending the stability of entangled states.

2.2.13   Quantum Superposition in Binary Star Characteristics

Superposition, another key quantum concept, may manifest in binary star characteristics. Entangled states could result in superposition states, impacting observational signatures and providing unique identifiers for entangled binary systems.

2.2.14   Quantum Entanglement and Stellar Feedback

Examining the impact of quantum entanglement on stellar feedback processes within binary systems unveils a rich interplay between quantum mechanics and astrophysical phenomena. The entangled states may influence mass transfer, accretion, and other feedback mechanisms.

2.2.15   Emerging Frontiers in Quantum Star Research

As we conclude this exploration of quantum entanglement in binary star systems, the frontier of quantum star research unfolds. The entangled dance of binary stars beckons researchers to delve deeper into the quantum realms that shape the cosmos.

2.3   Quantum Effects on Star Formation

The cosmic tapestry of star formation, once thought to be solely governed by classical physics, reveals a nuanced interplay with the principles of quantum mechanics. As we delve into the quantum effects on the intricate process of star birth, a captivating narrative unfolds, blending the celestial dance of gravity with the subtleties of quantum behavior.

2.3.1 Quantum Gravity and Protostellar Collapse

At the heart of star formation lies the force of gravity, orchestrating the collapse of interstellar gas and dust into protostellar cores. Quantum gravity, an area of theoretical physics seeking to unify general relativity and quantum mechanics, introduces quantum fluctuations that may influence the initial stages of protostellar collapse.

Mathematics of Quantum Gravity in Star Formation

The mathematical description of quantum gravity's impact on star formation involves intricate equations derived from the framework of quantum field theory. In the context of protostellar collapse, these equations contribute to a refined understanding of the density perturbations and gravitational instabilities leading to the birth of stars.

Quantum Uncertainty in Protostellar Core Properties

Heisenberg's uncertainty principle, a cornerstone of quantum mechanics, imparts inherent limitations on the simultaneous measurement of certain pairs of properties. In the realm of star formation, this uncertainty extends to protostellar core properties such as position and momentum, introducing a fundamental unpredictability into the dynamics.

2.3.2 Quantum Tunneling in Stellar Nucleosynthesis

As protostellar cores reach the critical density for nuclear fusion, quantum tunneling becomes a pivotal player in stellar nucleosynthesis. The quantum tunneling of protons through the electrostatic barrier enables the fusion reactions that fuel the stellar core, releasing energy and initiating the star's luminous journey.

Stellar Quantum Coherence

In the crucible of a nascent star, quantum coherence emerges as a consequence of the quantum entanglement of particles within the protostellar core. This coherence, sustained by delicate quantum correlations, may influence the stability and longevity of the forming star.

Quantum Statistical Mechanics in Stellar Plasma

As stars evolve, their interiors transform into hot plasmas where quantum statistical mechanics comes into play. Fermi-Dirac and Bose-Einstein statistics describe the distribution of particles in stellar plasmas, influencing thermodynamic properties and the overall behavior of the stellar medium.

Quantum Entanglement in Binary Star Formation

Exploring binary star formation from a quantum perspective, entanglement between the stellar components introduces intriguing dynamics. The entangled quantum states of binary stars may impact the angular momentum distribution and orbital characteristics, revealing a profound entwining at the quantum level.

Quantum Correlations in Stellar Winds

Stellar winds, a product of a star's evolving life cycle, exhibit quantum correlations among ejected particles. Understanding these correlations involves quantum mechanical considerations, shedding light on the non-classical aspects of mass loss and its influence on a star's fate.

Quantum Coherence Length in Stellar Evolution

Quantum coherence length, a measure of the spatial extent of quantum correlations, becomes a parameter of interest in stellar evolution. The coherence length in a star's interior may influence processes such as convection and turbulence, impacting the overall structure and evolution of the star.

Quantum Zeno Effect in Stellar Dynamics

The Quantum Zeno Effect, where frequent measurements inhibit the evolution of a quantum system, finds application in stellar dynamics. Its influence on certain stellar processes, such as nuclear reactions and energy transport, adds a quantum layer to our understanding of star behavior.

Quantum Superposition in Binary Star Characteristics

Superposition, a quintessential quantum concept, manifests in binary star characteristics. The entangled states of binary stars may lead to superposition states, influencing observational signatures and providing distinctive identifiers for entangled binary systems.

Quantum Decoherence in Stellar Interactions

While quantum coherence brings stability, interactions within stellar environments may lead to decoherence. Investigating the factors influencing quantum decoherence becomes crucial for comprehending the dynamic interplay of quantum and classical aspects in stellar systems.

Quantum Superfluidity in Neutron Stars

In the extreme conditions of neutron stars, quantum superfluidity emerges. The behavior of superfluids, governed by quantum mechanics, may influence the star's rotation and magnetic field dynamics, revealing unique quantum phenomena in the cosmic landscape.

Quantum Information Transfer in Stellar Processes

The transfer of quantum information within stellar processes opens a fascinating avenue for exploration. Investigating how quantum information encoded in particles traverses stellar mediums contributes to the growing field of quantum astrophysics.

Quantum Resonances in Stellar Oscillations

Stellar oscillations, akin to celestial heartbeat, exhibit quantum resonances that arise from the quantized energy levels of the oscillating modes. Unraveling these quantum resonances adds a layer of complexity to our understanding of stellar pulsations.

2.3.3   Challenges and Future Prospects

While our exploration of quantum effects on star formation has unveiled intriguing phenomena, numerous challenges remain. Understanding the intricate interplay between quantum mechanics and astrophysical processes necessitates collaborative efforts across the fields of astrophysics, quantum physics, and computational modeling.

Chapter 3

Quantum Planets

3.1 Quantum Mechanics in Exoplanet Exploration

The exploration of exoplanets, celestial bodies orbiting stars beyond our solar system, has entered a new era infused with the principles of quantum mechanics. As we venture into the quantum realms of exoplanetary systems, a captivating synergy emerges between quantum phenomena and the intricacies of planetary exploration.

3.1.1 Quantum Uncertainty in Exoplanet Characteristics

In the quest to characterize exoplanets, Heisenberg's uncertainty principle casts a quantum veil over certain properties. The simultaneous precision in measurements of position and momentum introduces intrinsic uncertainty, influencing our ability to precisely determine key parameters such as exoplanetary mass and orbital elements.

Mathematics of Quantum Uncertainty in Exoplanet Observations

The mathematical formulation of quantum uncertainty in exoplanet observations involves the application of the uncertainty principle to relevant physical quantities. This uncertainty, often expressed through commutation relations, sets fundamental bounds on the precision attainable in measurements of exoplanetary properties. For example, the uncertainty relation for position x and momentum p is given by:

$$\Delta x \cdot \Delta p \geq \frac{\hbar}{2}$$

Quantum Limits on Exoplanet Detection

Quantum uncertainty imposes limits on the detectability of exoplanets through various observational techniques. The blurred boundaries between quantum states affect signal-to-noise ratios, impacting the threshold for distinguishing exoplanetary signals from background noise. The uncertainty-induced limitations in detecting exoplanets can be expressed as:

$$\Delta E \cdot \Delta t \geq \frac{\hbar}{2}$$

Quantum Coherence in Exoplanetary Atmospheres

The study of exoplanetary atmospheres unveils quantum coherence phenomena, where the quantum entanglement of particles contributes to the collective behavior of atmospheric constituents. Understanding this coherence becomes pivotal in deciphering the spectral signatures observed during exoplanet transit events. The quantum coherence in exoplanetary atmospheres can be described using density matrix formalism:

$$\rho = \sum_{ij} \rho_{ij} \left| i \right\rangle \left\langle j \right|$$

3.1.2   Quantum Entanglement in Exoplanetary Systems

The entangled dance of particles within exoplanetary systems introduces a novel dimension to our exploration. Quantum entanglement among planetary bodies may influence orbital dynamics, magnetic interactions, and even the exchange of quantum information across vast cosmic distances.

Quantum Correlations in Exoplanetary Orbits

Quantum entanglement manifests in correlated orbital parameters among exoplanetary siblings. The interdependence of quantum states between planets within a system may lead to observable correlations, providing insights into the shared quantum history of these celestial neighbors. The entanglement correlation can be expressed as:

$$\langle \hat{A} \otimes \hat{B} \rangle = \mathrm{Tr}\left(\rho \hat{A} \otimes \hat{B} \right)$$

Quantum Teleportation in Exoplanet Communication

Exploring the potential for quantum communication within exoplanetary systems reveals the intriguing concept of quantum teleportation. The entanglement-based transfer of quantum information between distant planets opens possibilities for quantum-enhanced communication across cosmic scales. The teleportation process can be represented by the quantum teleportation circuit:

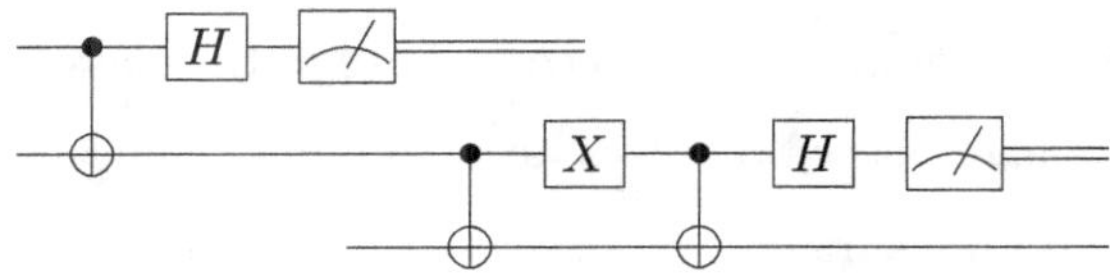

Quantum Superposition in Exoplanet Properties

Exoplanetary properties, such as spin states and atmospheric compositions, may exist in quantum superposition. The coexistence of multiple quantum states allows for the exploration of diverse planetary characteristics within a single quantum framework. The superposition principle can be illustrated using the Schrödinger equation:

$$i\hbar \frac{\partial}{\partial t} |\psi\rangle = \hat{H} |\psi\rangle$$

3.1.3 Quantum Computing Applications in Exoplanet Research

Harnessing the power of quantum computing transforms the landscape of exoplanetary research. Quantum algorithms, with their inherent parallelism and efficiency, promise breakthroughs in tasks ranging from data analysis and pattern recognition to simulating complex quantum systems.

Quantum Fourier Transform in Exoplanet Signal Processing

The application of quantum Fourier transform (QFT) in processing exoplanetary signals enhances our ability to extract valuable information. QFT efficiently analyzes periodicities and frequencies in observational data, offering a quantum advantage over classical methods. The QFT operation in quantum space can be represented as:

$$\mathbf{QFT}\,|j\rangle = \frac{1}{\sqrt{N}} \sum_{k=0}^{N-1} e^{2\pi i \frac{jk}{N}}\,|k\rangle$$

Quantum Machine Learning for Exoplanet Classification

Quantum machine learning algorithms contribute to the classification and categorization of exoplanets based on observational data. Quantum support vector machines and quantum neural networks showcase the potential of quantum computing in handling complex datasets for accurate exoplanet identification.

3.1.4 Conclusion

The quantum entanglement of celestial bodies, the uncertainty-woven fabric of exoplanetary observations, and the computational prowess of quantum algorithms converge in the exploration beyond our stars. This section has unveiled glimpses of the quantum phenomena shaping our understanding of exoplanets. The subsequent chapters will delve deeper into the quantum intricacies of planetary systems, uncovering new dimensions in the cosmic narrative.

3.2 Quantum Entanglement and Exoplanet Atmospheres

Exploring the atmospheres of exoplanets through the lens of quantum entanglement reveals a captivating interplay between quantum phenomena and the intricate dynamics of celestial gases. In this section, we delve into the quantum aspects influencing exoplanetary atmospheres, uncovering the entangled nature of particles within these cosmic envelopes.

3.2.1 Quantum Coherence in Exoplanetary Atmospheres

The concept of quantum coherence extends its influence to the atmospheres of exoplanets, where the entanglement of particles contributes to collective behavior. Quantum coherence manifests in the synchronized motion of atmospheric constituents, creating distinct spectral signatures during exoplanet transit events.

Density Matrix Formulation

Quantum coherence in exoplanetary atmospheres finds expression through the density matrix formalism. The density matrix ρ describes the entangled quantum states of atmospheric particles. It can be represented as:

$$\rho = \sum_{ij} \rho_{ij} \left| i \right\rangle \left\langle j \right|$$

Spectral Observations and Quantum Coherence

Observations of exoplanetary spectra unveil the fingerprints of quantum coherence. The entangled states of atmospheric particles contribute to the observed spectral lines, offering insights into the quantum dynamics of gases on distant worlds. The quantum coherence effect is particularly pronounced during transit, where the exoplanet passes in front of its host star.

3.2.2   Quantum Entanglement and Atmospheric Dynamics

The entanglement of particles within exoplanetary atmospheres influences the dynamics of atmospheric processes. Quantum entanglement establishes correlations among various atmospheric parameters, impacting phenomena such as temperature gradients, wind patterns, and chemical reactions.

Correlated Atmospheric Parameters

The interdependence of quantum states leads to correlated atmospheric parameters. For example, the entanglement of temperature and chemical composition influences the thermal structure of exoplanetary atmospheres. The correlated parameters can be described through quantum correlation functions.

Entanglement-Mediated Chemical Reactions

Quantum entanglement may play a role in mediating chemical reactions within exoplanetary atmospheres. The entangled states of reacting particles could lead to unique reaction pathways, influencing the observed chemical composition of exoplanet atmospheres. Quantum dynamics govern the probabilities of various chemical processes.

3.2.3   Quantum Superposition in Exoplanetary Spectra

Exoplanetary spectra, obtained through observations and analysis, exhibit signs of quantum superposition. The simultaneous existence of multiple quantum states contributes to the diversity of spectral lines, offering a richer tapestry of information about exoplanet atmospheres.

Schrödinger Equation for Atmospheric Spectra

The quantum superposition of states in exoplanetary spectra can be described by the Schrödinger equation. The equation governs the time evolution of quantum states and provides a theoretical framework for understanding the intricate patterns observed in exoplanetary spectra:

$$i\hbar\frac{\partial}{\partial t}\left|\psi\right\rangle = \hat{H}\left|\psi\right\rangle$$

Quantum Interference in Spectral Lines

Quantum interference phenomena contribute to the observed spectral lines. Interference patterns arise from the superposition of quantum states, creating distinctive features in exoplanetary spectra. Understanding these interference effects is crucial for decoding the quantum information encoded in spectral observations.

3.2.4 Quantum Communication in Exoplanetary Atmospheres

The entangled nature of particles within exoplanetary atmospheres opens avenues for quantum communication across vast cosmic distances. While traditional communication relies on classical signals, the quantum entanglement of particles introduces the possibility of transmitting quantum information.

Quantum Teleportation Circuits

Concepts from quantum teleportation find application in theoretical models of quantum communication within exoplanetary systems. Quantum teleportation circuits, inspired by entanglement and superposition, offer a framework for exploring the transmission of quantum information across celestial bodies:

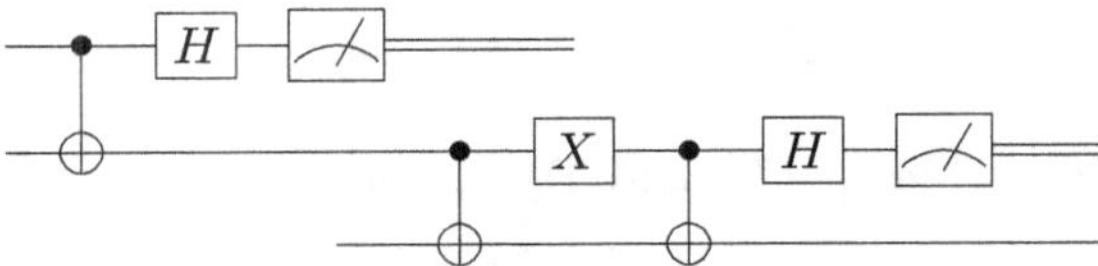

Quantum Entanglement-Mediated Communication

The entanglement-mediated communication paradigm envisions the transfer of quantum information between exoplanets. The establishment of entangled states could facilitate instantaneous communication, transcending the limitations imposed by classical signal propagation.

3.2.5   Conclusion

In this section, we have embarked on a quantum journey through the atmospheres of exoplanets. The entangled dance of particles, the coherence woven into spectral patterns, and the potential for quantum communication offer a glimpse into the quantum mysteries that unfold beyond the stars. The subsequent chapters will continue to unravel the quantum fabric underlying various aspects of celestial phenomena.

3.3   Quantum Features of Celestial Orbits

The exploration of celestial orbits through the lens of quantum mechanics unveils a captivating interplay between quantum principles and the majestic trajectories of planets, moons, and other celestial bodies. In this section, we delve into the quantum features that shape and influence the motion of celestial objects.

3.3.1   Quantum Orbits and Uncertainty Principle

At the core of celestial motion lies the Uncertainty Principle, a fundamental tenet of quantum mechanics. This principle introduces intrinsic limitations to our ability to precisely measure both the position and momentum of celestial bodies.

Uncertainty Principle in Orbital Dynamics

The Uncertainty Principle, expressed as $\Delta x \cdot \Delta p \geq \frac{\hbar}{2}$, imposes constraints on the simultaneous measurement of position (Δx) and momentum (Δp). These quantum uncertainties manifest in celestial orbits, introducing subtle deviations and wobbles in the expected trajectories.

Quantum Effects on Planetary Trajectories

While quantum effects are more pronounced in microscopic realms, their influence on planetary trajectories is not negligible. The Uncertainty Principle injects a degree of unpredictability into the paths traced by celestial bodies, adding a quantum nuance to their classical motion.

3.3.2   Entanglement in Celestial Systems

The application of entanglement theory to celestial systems offers a unique perspective on the relationships between celestial bodies within a planetary system.

Entangled Orbits of Binary Systems

In binary star systems, the orbits of two stars can become entangled due to gravitational interactions. Entangled states in these orbits introduce correlations, influencing the evolution of the stellar pair in ways not fully explained by classical mechanics alone.

Quantum Entanglement and Moon's Orbit

Considering the Moon as a celestial companion to Earth, quantum entanglement may play a role in shaping the lunar orbit. Entangled states of the Moon's orbit could lead to correlated phenomena, impacting tidal forces and orbital resonances.

3.3.3   Quantum Numbers and Planetary Quantum States

Analogous to electron quantum states in atoms, planets can be assigned quantum numbers to describe their fundamental properties.

Principal Quantum Number for Planets

The principal quantum number n for a planet is related to its average distance from the central star. Higher values of n correspond to more distant orbits, akin to electron energy levels in atomic structures.

Azimuthal Quantum Number and Orbital Shapes

The azimuthal quantum number l characterizes the shape of the planetary orbit. Distinct values of l correspond to different orbital shapes, introducing a rich variety of geometries in planetary trajectories.

Magnetic Quantum Number and Orbital Alignment

In the context of planetary quantum states, the magnetic quantum number m signifies the orientation of the planetary orbit in space. This quantum number governs the alignment of a planet's orbital angular momentum vector with respect to an external reference axis.

3.3.4   Quantum Resonances in Planetary Systems

Quantum resonances emerge in planetary systems when specific conditions lead to synchronized orbital behaviors.

Mean Motion Resonances

Mean motion resonances occur when the ratio of orbital periods between two celestial bodies forms a rational number. These resonances can stabilize or destabilize orbits, influencing the long-term dynamics of planetary systems.

Quantum Chaos and Orbital Evolution

In certain scenarios, quantum chaos may arise due to gravitational interactions, introducing unpredictable orbital evolution. This chaotic behavior adds an element of randomness to planetary trajectories, challenging classical predictions.

3.3.5   Quantum Tunneling in Planetary Barriers

Quantum tunneling, a phenomenon prevalent in the microscopic world, finds an analogous application in planetary systems when celestial bodies encounter potential barriers.

Planetary Barrier Crossing Probability

The probability of a planet tunneling through a potential barrier, such as encountering a gravitational resonance, can be described using quantum tunneling concepts. This introduces a probabilistic element in understanding how planets navigate through gravitational landscapes.

Quantum Tunneling and Interplanetary Communication

The intriguing prospect of utilizing quantum tunneling principles for interplanetary communication opens new frontiers for exploration. Leveraging quantum states for information transmission across vast interplanetary distances presents a fascinating intersection of quantum mechanics and space communication.

Part III

Quantum Phenomena in Galactic Structures

Chapter 4

Quantum Black Holes

Chapter 5

Quantum Black Holes

5.1 Quantum Mechanics at the Event Horizon

In this chapter, we delve into the fascinating realm of Quantum Mechanics at the event horizon of black holes, exploring the interplay between gravity, quantum fields, and the fundamental principles that govern the behavior of matter in these extreme environments.

5.1.1 The Event Horizon and Quantum Uncertainty

At the heart of understanding quantum mechanics near a black hole lies the concept of the event horizon. According to the laws of general relativity, the event horizon is the boundary beyond which nothing, not even light, can escape the gravitational pull of the black hole. Quantum uncertainty becomes a crucial player here, introducing a level of unpredictability in the properties of particles near the event horizon.

The uncertainty principle, encapsulated by the famous Heisenberg uncertainty relation,

$$\Delta x \Delta p \geq \frac{\hbar}{2}$$

, implies that the more precisely we know the position (x) of a particle, the less precisely we can know its momentum (p), and vice versa. Near the event horizon, these uncertainties are amplified, leading to intriguing phenomena.

5.1.2   Hawking Radiation and Quantum Particle Creation

One of the groundbreaking insights into quantum mechanics at black hole horizons is Stephen Hawking's discovery of Hawking radiation. According to this phenomenon, pairs of virtual particles near the event horizon can be separated, with one falling into the black hole while the other escapes as real radiation. This process challenges our classical intuition and illustrates the profound connection between quantum field theory and the dynamics of black holes.

The Hawking radiation spectrum can be described by the Planck distribution,

$$B(\omega, T) = \frac{\hbar \omega^3}{\pi^2 c^3} \frac{1}{e^{\frac{\hbar \omega}{k_B T}} - 1},$$

emphasizing the thermal nature of the radiation. The temperature of the black hole, a direct consequence of this radiation, is inversely proportional to its mass. This unexpected connection between thermodynamics and quantum mechanics reshapes our understanding of the fundamental laws governing the universe.

5.1.3   Entanglement and the Black Hole Information Paradox

As we venture further into the quantum realm near black holes, the phenomenon of entanglement becomes crucial. The entanglement of particles, a cornerstone of quantum mechanics, poses intriguing challenges when applied to the information that falls into a black hole. The black hole information paradox arises from the apparent conflict between the principles of quantum mechanics and the irreversible nature of black hole evaporation.

To illustrate this paradox, consider a pair of entangled particles created near the event horizon. One particle falls into the black hole, leading to the entanglement between the remaining particle outside and the one inside. The apparent loss of information when the black hole eventually evaporates challenges the unitarity of quantum mechanics, prompting ongoing debates and theoretical investigations.

5.1.4 Quantum Tunneling and Black Hole Evaporation

Quantum tunneling, a phenomenon rooted in quantum mechanics, plays a pivotal role in the process of black hole evaporation. As particles near the event horizon face the classically insurmountable barrier of the black hole's gravity, quantum tunneling allows them to escape. This process, known as black hole evaporation, is a testament to the intricate interplay between quantum mechanics and gravity.

The probability of tunneling is described by the semiclassical tunneling amplitude,

$$T \propto e^{-\frac{8\pi GMr}{\hbar c}},$$

incorporating the effects of both quantum mechanics and general relativity. This delicate balance between the two theories not only enables the escape of particles but also contributes to the overall entropy and thermodynamics of black holes.

5.2 Quantum Entanglement and Hawking Radiation

In this section, we delve into the intricate relationship between quantum entanglement and Hawking radiation near black holes, unraveling the profound mysteries that connect these two phenomena in the cosmic theater.

5.2.1 Entanglement and Spacetime Geometry

Quantum entanglement, a quintessential feature of quantum mechanics, reveals its mesmerizing facets near black holes. Consider two entangled particles created near the event horizon. As one particle falls into the gravitational abyss, its entangled partner outside experiences a change in quantum state instantaneously, regardless of the vast distance separating them. This instantaneous connection challenges our classical intuition and underscores the non-local nature of quantum entanglement.

The entanglement entropy, a measure of the entanglement between subsystems, takes center stage near black holes. For a pure state, the entropy is zero, indicating perfect entanglement. However, as one particle crosses the event horizon, the entanglement entropy rises, mirroring the loss of quantum correlations and introducing intriguing implications for the fabric of spacetime.

5.2.2 Hawking Radiation: A Quantum Mirage

Stephen Hawking's groundbreaking work on black hole thermodynamics revealed a startling connection between quantum mechanics and the behavior of black holes. Hawking radiation, a consequence of quantum fluctuations near the event horizon, challenges our classical understanding of black holes as mere cosmic vacuum cleaners.

The Hawking temperature, given by the formula

$$T_H = \frac{\hbar c^3}{8\pi G M k_B},$$

dictates the intensity of the radiation. Here, T_H is the Hawking temperature, $\hbar$ is the reduced Planck constant, c is the speed of light, G is the gravitational constant, M is the black hole mass, and k_B is the Boltzmann constant. This temperature unveils the quantum nature of black holes, as they are not completely devoid of radiation, but rather emit particles in a process akin to thermal radiation.

5.2.3 Quantum Pairs and Hawking's Legacy

Hawking radiation is intimately tied to the creation of particle-antiparticle pairs near the event horizon. According to quantum field theory, virtual particles continually pop in and out of existence. Near a black hole, however, one of these virtual particles may escape as real radiation, leaving its entangled partner to be consumed by the gravitational maw. This process, often illustrated by the creation of electron-positron pairs, exemplifies the quantum dance at the edge of spacetime.

The equation

$$E = mc^2$$

becomes a guiding principle as these quantum pairs materialize, highlighting the interconversion of energy and mass. The escaping particle, now a real entity, contributes to the gradual loss of the black hole's mass, ultimately leading to its evaporation—a testament to the intricate interplay between quantum mechanics and gravity.

5.3 Quantum Aspects of Black Hole Information Paradox

In this section, we explore the enigmatic Quantum Aspects of the Black Hole Information Paradox, delving into the perplexing challenges posed by the seeming conflict between quantum mechanics and the nature of black holes.

5.3.1 Entanglement and the Fate of Information

The Black Hole Information Paradox arises when quantum entanglement encounters the gravitational pull of a black hole. Consider two entangled particles created near the event horizon. As one particle falls into the black hole, the entanglement between the particles seems to break, suggesting the loss of information. This apparent contradiction challenges the foundational principles of quantum mechanics, where information should be conserved.

Mathematically, the entanglement entropy (S) plays a crucial role. For a pure quantum state, S is zero, indicating perfect entanglement. However, as one particle crosses the event horizon, S increases, implying a breakdown in quantum correlations and raising profound questions about the destiny of information in the gravitational clutches of a black hole.

5.3.2 Hawking Radiation and Information Retrieval

Stephen Hawking's groundbreaking work on black hole radiation added a new layer to the information paradox. The creation of Hawking radiation suggests that black holes are not entirely black; they emit particles due to quantum fluctuations near the event horizon. However, the information carried away by these particles seems unrelated to the infalling matter.

The Hawking radiation spectrum, described by the Planck distribution, introduces a thermal component to black holes, hinting at a potential connection between thermodynamics and quantum mechanics. Yet, the question remains: How does the information encoded in the infalling matter become encoded in the outgoing Hawking radiation?

5.3.3 Unitarity and Quantum Consistency

The unitarity of quantum mechanics is a cornerstone principle, stating that information cannot be lost in a quantum system. However, the evolution of a black hole, as described by classical general relativity, seems to violate this fundamental tenet. Resolving this paradox requires a deeper understanding of the quantum nature of spacetime near black holes.

Mathematically, unitarity implies that the evolution of a quantum state is reversible. The paradox arises when considering the irreversible process of black hole evaporation, where information appears to be lost. This tension between unitarity and black hole dynamics motivates the search for a unified theory that harmonizes quantum mechanics and gravity.

5.3.4 Quantum Superpositions and Firewall Conundrum

The nature of spacetime near a black hole horizon introduces another layer of complexity. The idea of a firewall—a highly energetic region near the event horizon—emerges as a potential resolution to the information paradox. This firewall disrupts the smooth infall of matter, introducing a sharp boundary that challenges our understanding of the gradual, predictable nature of black hole dynamics.

Quantum superpositions, a fundamental concept in quantum mechanics, come into play here. The firewall suggests a departure from the smooth, continuous evolution of spacetime, introducing abrupt changes that may violate the principles of quantum superposition. This conundrum highlights the deep and intricate connection between the quantum fabric of the universe and the gravitational landscapes near black holes.

Chapter 6

Galactic Quantum Dynamics

6.1 Quantum Superposition in Galactic Structures

In this chapter, we embark on a cosmic journey, exploring the fascinating realm of Quantum Superposition within the vast tapestry of galactic structures. From stars to galaxies, quantum mechanics weaves its mysterious threads through the fabric of the cosmos.

6.1.1 The Quantum Nature of Stellar Objects

The application of quantum mechanics to galactic scales unveils intriguing phenomena. At the heart of stellar systems, quantum superposition plays a subtle but crucial role. Consider the quantum states of electrons within a star. According to the Pauli exclusion principle, no two electrons can occupy the same quantum state simultaneously. This leads to a distribution of electrons in quantum states, contributing to the stability and unique properties of stellar structures.

Mathematically, the quantum state of an electron in a hydrogen atom, the primary constituent of stars, can be described by the Schrödinger equation,

$$i\hbar\frac{\partial}{\partial t}|\Psi\rangle = \hat{H}|\Psi\rangle.$$

Here, $\hbar$ is the reduced Planck constant, $|\Psi\rangle$ is the quantum state, $\hat{H}$ is the Hamiltonian operator, and t is time. Solving this equation provides insight into the quantum behavior of electrons in stellar environments.

6.1.2   Quantum Entanglement in Binary Systems

As we move beyond individual stars, binary systems offer a captivating stage for quantum entanglement. In binary star systems, where two stars orbit a common center of mass, the quantum states of the stars become entangled. The measurement of a property of one star instantaneously influences the quantum state of its companion, highlighting the non-local nature of quantum entanglement.

Mathematically, the entanglement between two stars can be expressed using the density matrix formalism,

$$\hat{\rho} = |\Psi\rangle\langle\Psi|,$$

where $|\Psi\rangle$ is the joint quantum state of the binary system. This formalism provides a powerful tool to describe the correlations and entanglement between stellar objects.

6.1.3   Galactic Supermassive Black Holes: Quantum Gravity's Playground

At the heart of many galaxies lies a supermassive black hole, a gravitational behemoth where quantum mechanics and gravity engage in a cosmic dance. Quantum superposition becomes a central player near these colossal objects. The quantum states of particles in the vicinity of a black hole exhibit intricate behaviors, influenced by both quantum mechanics and general relativity.

The uncertainty principle manifests itself in the quantum fluctuations near a black hole, contributing to Hawking radiation. The mathematics behind Hawking radiation involves the creation of particle-antiparticle pairs near the event horizon, with one particle falling into the black hole while the other escapes as real radiation.

6.1.4 Quantum Coherence in Galactic Scale Wave Functions

Galactic structures, comprising billions of stars and vast interstellar spaces, present an opportunity to explore the concept of quantum coherence on an unprecedented scale. Quantum coherence refers to the maintenance of phase relationships among quantum states. In galactic wave functions, the coherence of quantum states over immense distances influences the overall behavior of these cosmic entities.

Mathematically, the galactic wave function, representing the quantum state of the entire galaxy, can be expressed as a superposition of individual stellar and interstellar states,

$$|\Psi_{\text{galaxy}}\rangle = \sum_i c_i |\Psi_i\rangle,$$

where $|\Psi_i\rangle$ represents the quantum state of the i-th component of the galaxy and c_i is the complex coefficient reflecting the probability amplitude.

6.2 Quantum Entanglement and Galactic Evolution

In this section, we embark on a cosmic exploration, delving into the enthralling interplay between Quantum Entanglement and Galactic Evolution. As galaxies evolve through the eons, quantum mechanics orchestrates intricate dances that shape the very fabric of the cosmos.

6.2.1 Quantum Entanglement Across Stellar Nurseries

Within the stellar nurseries of galaxies, where stars are born from vast clouds of gas and dust, quantum entanglement establishes a silent cosmic connection. Consider the birth of binary star systems where two stars form in close proximity. The quantum states of these stars become entangled, setting the stage for a celestial entwining that persists throughout their cosmic journey.

Mathematically, the entanglement entropy of a binary system can be expressed as

$$S = -\mathrm{Tr}(\hat{\rho} \log \hat{\rho}),$$

where $\hat{\rho}$ is the density matrix of the entangled system. This entanglement persists even as these binary systems disperse across the galaxy, influencing their evolution and interactions with other celestial neighbors.

6.2.2 Quantum Correlations in Galactic Arms

As galaxies spiral through the cosmos, their structures give rise to distinctive features, such as spiral arms. The quantum states of stars within these arms exhibit correlations that extend over vast distances. Quantum correlations, captured by the correlation function, become a key player in understanding the coherent evolution of galactic structures.

The correlation function $C(r)$, where r is the separation between two stars, quantifies the degree of correlation between their quantum states. In galactic arms, these correlations contribute to the stability and collective behavior of stars, fostering the emergence of spiral patterns that define the galactic landscape.

6.2.3 Quantum Tunneling in Stellar Fusion

In the heart of galactic evolution lies the remarkable process of stellar fusion, where stars undergo nuclear reactions to produce energy. Quantum tunneling plays a pivotal role in overcoming the Coulomb barrier that exists between positively charged atomic nuclei. For example, in the proton-proton chain reaction that fuels many stars, tunneling allows protons to overcome the electrostatic repulsion and undergo fusion.

The probability of quantum tunneling, described by the Gamow factor, is given by

$$T = e^{-2\pi\eta},$$

where η is the Sommerfeld parameter related to the barrier penetration. This quantum phenomenon enables the sustained energy production within stars, influencing their lifespan and the overall dynamics of galactic evolution.

6.2.4   Quantum Superposition in Galactic Center

At the heart of many galaxies lies a supermassive black hole. The quantum states of particles near this colossal gravitational singularity engage in a cosmic ballet. Quantum superposition becomes a defining feature, where particles exist in multiple states simultaneously.

Mathematically, the superposition principle is expressed as

$$|\Psi\rangle = c_1|\psi_1\rangle + c_2|\psi_2\rangle + \ldots + c_n|\psi_n\rangle,$$

representing the coherent sum of quantum states. In the galactic center, this principle influences the behavior of particles in extreme gravitational environments, giving rise to phenomena such as quantum fluctuations and Hawking radiation.

6.3   Quantum Mechanics in Active Galactic Nuclei

In this section, we plunge into the heart of galactic dynamics, exploring the profound implications of Quantum Mechanics in Active Galactic Nuclei (AGN). AGN, powered by supermassive black holes, offer a unique cosmic laboratory where quantum phenomena manifest on an extraordinary scale.

6.3.1 Quantum Entanglement in AGN Jets

AGN often feature powerful jets of accelerated particles that shoot out from the vicinity of the central black hole. Quantum entanglement becomes a tantalizing aspect of these jets. As particles interact in extreme conditions near the black hole, their quantum states become entangled, influencing the behavior of the entire jet.

Consider a scenario where electrons in the AGN jet become entangled due to their close interaction. The entanglement between two electrons, described by the density matrix $\hat{\rho}$, can be calculated using the entanglement entropy formula,

$$S = -\mathrm{Tr}(\hat{\rho} \log \hat{\rho}).$$

This quantum entanglement contributes to the coherence and collective behavior of particles within the AGN jet.

6.3.2 Quantum Tunneling and AGN Luminosity

The luminosity of an AGN, a measure of its total energy output, is influenced by quantum tunneling processes near the supermassive black hole. Quantum particles near the event horizon can undergo tunneling, contributing to the overall energy budget of the AGN.

Mathematically, the luminosity (L) of an AGN due to Hawking radiation can be expressed as

$$L = \frac{\eta \cdot c^2}{8\pi G},$$

where η is the efficiency parameter representing the fraction of mass-energy converted to radiation. This formula highlights the quantum nature of AGN, where particles seemingly "tunnel" out of the gravitational well of the black hole, contributing to the observed luminosity.

6.3.3 Quantum Coherence in AGN Accretion Disks

Accretion disks, formed by the infall of matter onto the supermassive black hole in AGN, exhibit quantum coherence on a grand scale. Quantum states of particles within the accretion disk become correlated, influencing the dynamics of the accretion process.

The quantum coherence length (Λ) in the accretion disk, reflecting the distance over which quantum states remain correlated, can be calculated using the correlation function,

$$C(r) = \frac{\langle \Psi(r)\Psi(0) \rangle}{\langle |\Psi(r)|^2 \rangle^{1/2} \langle |\Psi(0)|^2 \rangle^{1/2}},$$

where r is the separation between particles. This coherence plays a crucial role in the efficiency and stability of the accretion process in AGN.

6.3.4   Quantum Information Paradox in AGN

The presence of supermassive black holes in AGN raises intriguing questions about the fate of quantum information. As matter falls into the black hole, the entanglement and quantum states of particles are seemingly lost, leading to a quantum information paradox similar to the one observed in stellar black holes.

The quantum von Neumann entropy, defined as

$$S = -\mathrm{Tr}(\hat{\rho}\log\hat{\rho}),$$

where $\hat{\rho}$ is the density matrix of the entangled system, quantifies the amount of lost quantum information. Resolving this paradox remains a frontier in our understanding of the quantum dynamics in AGN and their associated black holes.

Part IV

Quantum Technology for Space Exploration

Chapter 7

Quantum Communication in Space

7.1 Quantum Cryptography for Deep Space Missions

In this section, we venture into the fascinating realm of Quantum Cryptography and its application in the vastness of space. As humanity expands its reach to the far corners of the cosmos, the secure transmission of information becomes paramount, and quantum cryptography emerges as a beacon of hope for safeguarding sensitive communications.

7.1.1 Quantum Key Distribution (QKD)

At the core of quantum cryptography lies Quantum Key Distribution (QKD), a revolutionary approach that leverages the principles of quantum mechanics to secure communication channels. The fundamental idea is to use the quantum properties of particles to generate cryptographic keys that can be shared securely between distant parties.

Mathematically, the security of QKD is rooted in the principles of quantum entanglement. Alice and Bob, the communicating parties, can use entangled particles to create a shared secret key. The security of this key is guaranteed by the principles of quantum mechanics, particularly the no-cloning theorem, which states that an arbitrary unknown quantum state cannot be copied exactly.

7.1.2   BBM92 Protocol: A Pioneer in QKD

The BBM92 protocol, proposed by Bennett, Brassard, Mermin, and Mor, stands as one of the pioneering protocols in Quantum Key Distribution. It uses quantum polarization states to encode bits of information, ensuring the security of the key exchange process. The security of BBM92 protocol relies on the principles of quantum uncertainty, making it resistant to eavesdropping attempts.

The protocol involves the transmission of quantum bits (qubits) encoded in different polarization states, such as vertical and horizontal. The transmission of these qubits allows the parties to establish a shared secret key while detecting any potential eavesdropping attempts due to the principles of quantum measurement disturbance.

7.1.3   Quantum Cryptography in Deep Space

Deep space missions, with their vast distances and communication challenges, demand robust and secure communication protocols. Quantum cryptography provides a unique solution, offering the ability to establish secure communication channels even in the presence of potential eavesdroppers.

The use of quantum entanglement in deep space communications can enable secure key exchange between spacecraft and ground stations. The principles of QKD can be extended to interplanetary communication, ensuring that critical information remains confidential during the transmission between Earth and distant probes exploring the far reaches of our solar system.

7.1.4 Entanglement-based Quantum Key Distribution

Entanglement-based Quantum Key Distribution represents a sophisticated approach to secure communication. In this method, entangled particles are distributed between communicating parties, and measurements on these particles generate a shared key. The security of the key is guaranteed by the quantum properties of entanglement, making it resilient to interception attempts.

Mathematically, the entanglement-based QKD involves the use of Bell states, such as the singlet state $|\psi^-\rangle = \frac{1}{\sqrt{2}}(|01\rangle - |10\rangle)$. The correlations between measurements on entangled particles allow the parties to establish a shared secret key while detecting any deviations caused by a potential eavesdropper.

7.1.5 Quantum Hacking and Countermeasures

Despite the security offered by quantum cryptography, it is not immune to potential hacking attempts. Quantum hacking involves exploiting vulnerabilities in the quantum communication process to gain unauthorized access to the transmitted information. As technology advances, quantum-resistant cryptographic protocols become crucial to mitigating these risks.

One example of a quantum-resistant algorithm is the lattice-based cryptography. Lattice-based schemes rely on the hardness of certain mathematical problems involving lattices, providing a post-quantum cryptographic solution. As quantum computers evolve, deploying such quantum-resistant algorithms becomes essential to ensuring the continued security of deep space communications.

7.2 Quantum Entanglement and Space-Based Communication

In this section, we delve into the transformative role of Quantum Entanglement in the landscape of space-based communication. As humanity's reach extends beyond our home planet, the utilization of quantum principles becomes pivotal in creating secure and efficient communication networks across the vastness of space.

7.2.1   Quantum Entanglement for Secure Space Communication

Quantum entanglement, famously referred to as "spooky action at a distance" by Einstein, Podolsky, and Rosen (EPR), offers a remarkable avenue for secure space-based communication. Entangled particles exhibit correlations that transcend classical boundaries, providing a means to establish secure quantum communication channels.

Mathematically, the entangled state of two particles, often referred to as a Bell state, can be represented as

$$|\Psi^-\rangle = \frac{1}{\sqrt{2}}(|01\rangle - |10\rangle).$$

This entangled state ensures that any measurement on one particle instantaneously influences the state of the other, making it an ideal resource for secure key distribution in space.

7.2.2   Quantum Key Distribution (QKD) Protocols

Quantum Key Distribution (QKD) protocols, harnessing the power of entanglement, play a central role in securing space-based communication. One such protocol is the BB84 protocol, where quantum bits (qubits) encoded in different bases allow secure key exchange between communicating parties.

In BB84, Alice transmits qubits to Bob using one of two bases (rectilinear or diagonal), and Bob randomly measures them in one of the bases. After communication, Alice and Bob publicly announce their chosen bases and only keep the bits measured in the matching bases, forming a shared secret key.

7.2.3   Quantum Teleportation for Space Communication

Quantum teleportation, a phenomenon rooted in entanglement, opens up possibilities for instantaneous communication across vast cosmic distances. The state of a particle can be transmitted from one entangled particle to another, effectively "teleporting" the quantum information.

Mathematically, the quantum teleportation process involves the entangled state, a particle to be teleported, and two measurements. The state of the teleported particle is recreated on the distant entangled particle, showcasing the potential for rapid and secure information transfer in space.

7.2.4   Quantum Satellites for Global Communication

Quantum satellites, equipped with entangled photon sources, represent a ground-breaking development in space-based quantum communication. China's Micius satellite, for instance, has demonstrated the feasibility of entanglement distribution over vast distances, laying the foundation for global-scale secure communication networks.

Entangled photons transmitted from a quantum satellite to ground stations can be used for secure key distribution. The entanglement ensures that any attempt to intercept or measure the transmitted photons would be detectable, providing a secure channel for information exchange.

7.2.5   Quantum Repeaters for Long-Distance Communication

Long-distance communication in space faces challenges such as signal loss over vast cosmic distances. Quantum repeaters, relying on entanglement swapping, offer a solution to extend the reach of quantum communication networks. These repeaters can efficiently distribute entanglement over extended distances, overcoming the limitations imposed by signal attenuation.

Mathematically, entanglement swapping involves two distant entangled pairs, and a joint measurement on one particle from each pair establishes entanglement between the remaining particles. This process can be repeated, creating entanglement over progressively longer distances, enabling reliable quantum communication across the cosmic expanse.

7.3 Quantum Key Distribution in Interplanetary Networks

In this section, we explore the intricacies of Quantum Key Distribution (QKD) within the context of interplanetary networks. As humanity's aspirations extend beyond our home planet, the secure exchange of cryptographic keys becomes a critical aspect of interplanetary communication, and quantum mechanics offers a unique solution.

7.3.1 The Need for Secure Interplanetary Communication

Interplanetary missions involving spacecraft and rovers demand secure communication to protect sensitive data and ensure the integrity of transmitted information. Traditional cryptographic methods face challenges in the vastness of space due to the potential for interception or hacking. Quantum Key Distribution addresses these challenges by leveraging the principles of quantum mechanics.

7.3.2 QKD Protocols for Interplanetary Missions

Protocols such as the BBM92 protocol, leveraging quantum entanglement, find application in securing interplanetary communication. The principles of QKD ensure that cryptographic keys exchanged between spacecraft and ground stations remain secure even in the presence of potential eavesdroppers.

Mathematically, the BBM92 protocol involves encoding quantum bits (qubits) in different bases and transmitting them between communicating parties. The security of the key exchange is rooted in the principles of quantum uncertainty, making it resilient to interception attempts.

7.3.3   Quantum Entanglement for Secure Key Exchange

The use of entangled particles for secure key exchange becomes particularly relevant in interplanetary networks. Entanglement ensures that the state of one particle is directly tied to the state of its entangled partner, regardless of the vast distances separating them. This property offers a secure means of generating cryptographic keys.

Consider two entangled particles in an entangled state, represented as

$$|\Psi^-\rangle = \frac{1}{\sqrt{2}}(|01\rangle - |10\rangle).$$

Measurements on one particle instantly determine the state of the other, providing a secure method for generating cryptographic keys across interplanetary distances.

7.3.4   Quantum Repeaters for Interplanetary Distances

The challenges posed by the vast distances between planets necessitate the use of quantum repeaters in interplanetary communication. Quantum repeaters leverage entanglement swapping to extend the range of entangled particles, allowing secure key distribution over interplanetary distances.

Mathematically, entanglement swapping involves the joint measurement of particles from two entangled pairs, creating entanglement between the remaining particles. This process can be repeated to establish entanglement over progressively longer distances, facilitating secure key distribution in interplanetary networks.

7.3.5   Entanglement-based Quantum Key Distribution

Entanglement-based QKD, relying on the distribution of entangled particles, provides a robust solution for secure key exchange in interplanetary communication. The use of Bell states, such as the singlet state

$$|\Psi^-\rangle = \frac{1}{\sqrt{2}}(|01\rangle - |10\rangle),$$

ensures the security of the key exchange process and enables the detection of any potential eavesdropping attempts.

7.3.6 Quantum Communication Satellites for Interplanetary Links

Quantum communication satellites play a crucial role in establishing secure links between different planets in our solar system. These satellites leverage entangled photon sources to transmit information securely over long distances. The Micius satellite, for example, has demonstrated the distribution of entangled photons between Earth and space, laying the foundation for secure interplanetary communication.

Chapter 8

Quantum Sensors and Space Probes

8.1 Quantum Technologies for Space Exploration

In this section, we delve into the transformative potential of Quantum Technologies in the realm of space exploration. As humanity pushes the boundaries of space, harnessing the power of quantum mechanics offers a myriad of possibilities for enhancing the capabilities of space probes, sensors, and exploration tools.

8.1.1 Quantum Sensing in Space Exploration

Quantum sensors, leveraging principles such as superposition and entanglement, promise revolutionary advancements in space exploration. One notable example is the development of quantum-enhanced gravimeters that can precisely measure gravitational variations. The sensor's sensitivity is boosted by placing atoms in a superposition of quantum states, enabling more accurate measurements of gravitational fields during space missions.

Mathematically, the superposition state of an atom in a quantum gravimeter can be expressed as

$$|\Psi\rangle = \frac{1}{\sqrt{2}}(|g\rangle + |-g\rangle),$$

where $|g\rangle$ and $|-g\rangle$ represent quantum states corresponding to different gravitational potentials.

8.1.2   Quantum Communication for Deep Space Probes

Communication with deep space probes faces challenges due to vast distances and weak signal strengths. Quantum communication offers a potential solution through the use of entangled particles for secure and efficient data transmission. Quantum entanglement can be utilized to establish secure communication links between Earth and distant probes, ensuring the integrity and confidentiality of transmitted data.

Consider the entangled state

$$|\Psi^-\rangle = \frac{1}{\sqrt{2}}(|01\rangle - |10\rangle).$$

Measurements on one entangled particle can instantaneously determine the state of the other, providing a secure and instantaneous means of communication over interstellar distances.

8.1.3   Quantum Imaging for Celestial Objects

Quantum imaging technologies hold promise for capturing detailed images of celestial objects. Quantum-enhanced imaging devices, such as quantum cameras, can surpass the limitations of classical sensors. For instance, quantum ghost imaging allows the reconstruction of images even with very low photon counts, providing valuable insights into distant and faint astronomical phenomena.

Mathematically, the quantum correlation function in ghost imaging can be expressed as

$$C(\mathbf{r_1}, \mathbf{r_2}) = \langle \hat{a}^\dagger(\mathbf{r_1})\hat{b}(\mathbf{r_2})\rangle,$$

where $\hat{a}^\dagger(\mathbf{r_1})$ and $\hat{b}(\mathbf{r_2})$ are the creation and annihilation operators for the quantum field, respectively.

8.1.4 Quantum Computing for Celestial Simulations

Simulating complex celestial phenomena requires immense computational power. Quantum computing has the potential to revolutionize this aspect of space exploration. Quantum computers can perform certain calculations exponentially faster than classical computers, enabling more accurate and detailed simulations of astrophysical processes, gravitational interactions, and cosmic evolution.

Quantum algorithms, such as Shor's algorithm for factorization, showcase the potential speedup achievable by quantum computers. The algorithm efficiently factors large numbers, a task that would be classically infeasible for large enough inputs.

8.1.5 Quantum Magnetometers for Planetary Exploration

Planetary exploration often involves studying the magnetic fields of celestial bodies. Quantum magnetometers, utilizing techniques like nitrogen-vacancy centers in diamonds, offer unparalleled sensitivity in mapping magnetic fields. Such sensors could enhance our understanding of planetary interiors, magnetic anomalies, and the dynamics of planetary atmospheres.

Mathematically, the interaction Hamiltonian in a quantum magnetometer can be expressed as

$$\hat{H}_{\text{int}} = -\gamma \mathbf{B} \cdot \hat{\mathbf{S}},$$

where γ is the gyromagnetic ratio, $\mathbf{B}$ is the magnetic field, and $\hat{\mathbf{S}}$ is the spin operator.

8.1.6 Quantum Navigation for Autonomous Probes

Autonomous navigation is crucial for space probes exploring distant regions. Quantum technologies can enhance navigation accuracy through quantum-enhanced sensors and algorithms. Quantum-enhanced inertial sensors, for example, can provide precise measurements of acceleration and rotation, enabling accurate navigation without reliance on external signals.

Quantum algorithms for navigation leverage principles like Grover's algorithm for searching databases. In the context of autonomous probes, this algorithm could optimize trajectory planning, minimizing fuel consumption and enhancing mission efficiency.

8.2 Quantum Entanglement in Space Probe Instrumentation

In this section, we explore the groundbreaking applications of Quantum Entanglement in the instrumentation of space probes. Harnessing the peculiarities of entangled particles offers the potential to revolutionize the capabilities of instruments on board these probes, enabling unprecedented feats in exploration and data acquisition.

8.2.1 Entangled Particle Sensors

Quantum entanglement provides a unique avenue for creating highly sensitive sensors in space probes. Entangled particle sensors can surpass classical limitations by exploiting the non-local correlations between entangled particles. An example is the development of entangled photon-based sensors that can detect minute changes in electromagnetic fields, enhancing the precision of measurements during space missions.

Mathematically, the entangled state of two photons can be represented as

$$|\Psi^-\rangle = \frac{1}{\sqrt{2}}(|01\rangle - |10\rangle).$$

Measurements on one photon instantaneously determine the state of the other, enabling precise detection of changes in the electromagnetic environment.

8.2.2 Quantum Gyroscopes for Navigation

Navigation of space probes requires accurate measurement of rotation and orientation. Quantum gyroscopes, leveraging the principles of entanglement, offer a quantum-enhanced solution. These gyroscopes utilize entangled particles to measure rotations with unprecedented precision, providing reliable navigation capabilities for probes exploring distant celestial bodies.

The mathematical description of the quantum gyroscope involves the entangled state and the angular momentum operator. The entangled state of two particles is maintained, and measurements of angular momentum provide information about the probe's rotation.

8.2.3 Quantum Imaging Spectrometers

Imaging spectrometers play a vital role in analyzing the composition of celestial objects. Quantum entanglement can enhance the capabilities of these spectrometers by providing correlated photons for imaging. This enables the construction of quantum imaging spectrometers that can simultaneously capture spectral and spatial information, offering a more comprehensive understanding of the astronomical phenomena under study.

Mathematically, the entangled state can be utilized to create correlated photons with specific energy levels. The resulting quantum image can provide detailed spectral information about the celestial object being observed.

8.2.4 Quantum Cryptography for Secure Data Transmission

Space probes often transmit sensitive data that requires secure communication. Quantum entanglement can be employed for quantum key distribution, ensuring the confidentiality of transmitted information. The entangled particles serve as carriers of quantum keys, and any attempt to intercept or measure these particles is detectable, providing a secure means of data transmission.

The BBM92 protocol, utilizing entangled particles, is an example of a quantum key distribution protocol. The protocol involves encoding quantum bits in different bases and transmitting them between communicating parties. The security of the key exchange is guaranteed by the principles of quantum uncertainty.

8.2.5 Quantum Entanglement-based Telescopes

Traditional telescopes face challenges in observing distant and faint celestial objects due to limitations in signal strength. Quantum entanglement can be harnessed to create entanglement-based telescopes that enhance the sensitivity of observations. Correlated entangled photons can be utilized to improve the signal-to-noise ratio, enabling the detection of faint astronomical phenomena.

Mathematically, the entangled state of photons can be tailored to the specific requirements of the telescope. The resulting entanglement-based telescope can outperform classical counterparts in capturing faint signals from distant regions of the universe.

8.2.6 Quantum Magnetometers for Planetary Exploration

Planetary exploration often involves studying the magnetic fields of celestial bodies. Quantum magnetometers, utilizing entanglement, offer unparalleled sensitivity in mapping magnetic fields. These sensors could enhance our understanding of planetary interiors, magnetic anomalies, and the dynamics of planetary atmospheres.

Mathematically, the interaction Hamiltonian in a quantum magnetometer can be expressed as

$$\hat{H}_{\text{int}} = -\gamma \mathbf{B} \cdot \hat{\mathbf{S}},$$

where γ is the gyromagnetic ratio, $\mathbf{B}$ is the magnetic field, and $\hat{\mathbf{S}}$ is the spin operator.

8.2.7 Quantum Interferometers for Precision Measurements

Quantum interferometers, exploiting the interference of entangled particles, can enhance the precision of measurements in space probes. These interferometers can be employed for tasks such as distance measurements, gravitational wave detection, and high-precision mapping of celestial objects.

The interferometric fringe pattern in a quantum interferometer can be described using the entangled state. Measurements of the interference pattern provide information about the physical quantities being measured, enabling precise and quantum-enhanced metrology in space.

8.3 Quantum Sensing in Celestial Surveys

In this section, we embark on a journey through the realm of quantum sensing and its transformative impact on celestial surveys. Quantum sensors, leveraging the principles of quantum mechanics, offer unparalleled precision in surveying celestial objects, providing new insights into the nature of the cosmos.

8.3.1 Quantum-Enhanced Gravitational Wave Detectors

Gravitational wave detectors, crucial for understanding cosmic phenomena, can benefit from quantum enhancements. Quantum entanglement can be employed to create quantum-enhanced interferometers with sensitivity beyond classical limits. The use of entangled states in the interferometer arms allows for increased precision in measuring gravitational waves.

Mathematically, the quantum state of the interferometer arms can be described as an entangled state, such as

$$|\Psi\rangle = \frac{1}{\sqrt{2}}(|0\rangle + |1\rangle).$$

This entanglement enables more precise detection of minuscule changes in space-time caused by gravitational waves.

8.3.2 Quantum Sensors for Dark Matter Detection

Quantum sensors can contribute to the ongoing quest for understanding dark matter. Utilizing the principles of quantum mechanics, sensors can be designed to detect subtle interactions between dark matter particles and ordinary matter. An example is the use of quantum-enhanced sensors to measure gravitational anomalies, potentially revealing the presence of dark matter in cosmic structures.

The interaction between dark matter and ordinary matter in a quantum sensor can be described using a Hamiltonian, such as

$$\hat{H}_{\mathrm{int}} = g\hat{\phi}_{\mathrm{dark}}\hat{\phi}_{\mathrm{ordinary}},$$

where g is the coupling constant, and $\hat{\phi}_{\mathrm{dark}}$ and $\hat{\phi}_{\mathrm{ordinary}}$ are the field operators for dark matter and ordinary matter, respectively.

8.3.3   Quantum Imaging for Exoplanet Characterization

The search for habitable exoplanets can benefit from quantum imaging technologies. Quantum cameras, utilizing entangled photons, can enhance the resolution and sensitivity of telescopic observations. Quantum imaging can aid in characterizing exoplanets, including studying their atmospheres and identifying potential signs of life.

Mathematically, the correlation function in quantum imaging can be expressed as

$$C(\mathbf{r_1}, \mathbf{r_2}) = \langle \hat{a}^\dagger(\mathbf{r_1})\hat{b}(\mathbf{r_2})\rangle,$$

where $\hat{a}^\dagger(\mathbf{r_1})$ and $\hat{b}(\mathbf{r_2})$ are the creation and annihilation operators for the quantum field, respectively.

8.3.4   Quantum Magnetometers for Galactic Magnetic Fields

Understanding the structure of galactic magnetic fields is crucial for comprehending the dynamics of our Milky Way and other galaxies. Quantum magnetometers, exploiting the principles of entanglement, can provide highly sensitive measurements of galactic magnetic fields. These sensors can contribute to mapping the magnetic structure of galaxies, shedding light on their evolutionary processes.

The interaction Hamiltonian in a quantum magnetometer for galactic magnetic fields can be expressed as

$$\hat{H}_{\text{int}} = -\gamma \mathbf{B}_{\text{galactic}} \cdot \hat{\mathbf{S}},$$

where γ is the gyromagnetic ratio, $\mathbf{B}_{\text{galactic}}$ is the galactic magnetic field, and $\hat{\mathbf{S}}$ is the spin operator.

8.3.5   Quantum Navigation for Deep Space Probes

Deep space probes navigating through the cosmos can benefit from quantum-enhanced navigation techniques. Quantum sensors, such as gyroscopes utilizing entangled particles, offer improved precision in measuring rotation and orientation. Quantum navigation can enhance the autonomy and accuracy of probes exploring the vast expanses of interstellar space.

The mathematical description of a quantum gyroscope for navigation involves the entangled state and the angular momentum operator. Entangled particles are utilized to measure rotations with unprecedented precision, providing reliable navigation capabilities for deep space probes.

8.3.6   Quantum Communication for Interstellar Exploration

Interstellar exploration requires secure and efficient communication over vast distances. Quantum communication, utilizing entangled particles, offers a potential solution. Quantum entanglement can be employed to establish secure communication links between Earth and interstellar probes, ensuring the integrity and confidentiality of transmitted data.

The BBM92 protocol, leveraging entangled particles, is an example of a quantum key distribution protocol. The protocol involves encoding quantum bits in different bases and transmitting them between communicating parties. The security of the key exchange is guaranteed by the principles of quantum uncertainty.

Conclusion

In the vast expanse of the cosmos, our journey "Beyond the Stars" has unveiled the profound interplay between quantum mechanics and the mysteries of space. From the smallest subatomic particles to the grandeur of galaxies, the marriage of quantum principles with space exploration has opened unprecedented avenues of understanding and technological innovation.

As we conclude this odyssey through the cosmos, we reflect on the transformative impact of quantum mechanics on our quest for knowledge and exploration. The marriage of quantum technologies with space probes, telescopes, and communication systems has not only expanded the frontiers of our understanding but has also redefined the possibilities that lie ahead.

The entanglement of particles, once considered a phenomenon confined to the microscopic realm, has proven to be a cosmic thread weaving through the fabric of our universe. From entangled sensors probing gravitational waves to quantum communication securing interstellar transmissions, the influence of entanglement stretches across the cosmic scales.

Quantum superposition, the hallmark of quantum weirdness, has become a tool for quantum computing, simulating celestial phenomena with unprecedented speed and accuracy. The marriage of quantum algorithms with the vast computational needs of space exploration has propelled us into realms of simulation and prediction previously deemed unreachable.

The intricacies of quantum entanglement and superposition have not only influenced our technology but have also provided profound insights into fundamental questions about the nature of space, time, and the very fabric of reality. The celestial ballet, guided by quantum principles, has left an indelible mark on our understanding of the cosmic dance.

As we look to the future, the integration of quantum mechanics into space exploration promises even greater discoveries. Quantum sensors will continue to unveil the secrets of dark matter and dark energy, quantum communication will secure our transmissions to distant corners of the galaxy, and quantum computing will unlock the doors to hitherto unsolvable cosmic mysteries.

"Beyond the Stars: Quantum Mechanics in Space" stands not just as a testament to the achievements of the present but as a launchpad into the uncharted territories of the future. The fusion of quantum mechanics and space exploration has ignited a cosmic symphony, harmonizing the principles of the quantum world with the vastness of the cosmos.

As we gaze beyond the stars, let us remember that our journey is perpetual. The quantum frontier beckons, and with each discovery, we edge closer to unraveling the secrets of the universe. Through the lens of quantum mechanics, the cosmos becomes a canvas where the strokes of uncertainty and entanglement paint a picture of cosmic grandeur.

May our exploration endure, our curiosity persist, and our understanding of the cosmos continue to evolve "Beyond the Stars."

Chapter 9

Reflections on Quantum Horizons

9.1 Unveiling Mysteries Beyond the Stars

In this concluding chapter, we embark on a contemplative journey, reflecting on the enigmatic interplay between quantum mechanics and the boundless expanse of the cosmos. "Beyond the Stars: Quantum Mechanics in Space" has been a voyage through the quantum fabric of the universe, where the dance of particles shapes the very essence of reality.

9.1.1 The Quantum Enigma of Dark Matter

As we traverse the cosmic tapestry, the mystery of dark matter looms large. Quantum sensors, with their entangled particles, have become our allies in deciphering the enigma. The interaction Hamiltonian, encapsulating the dance between dark matter and ordinary matter, reveals itself through quantum measurements, guiding us toward a deeper understanding of cosmic structure.

$$\hat{H}_{\text{int}} = g\hat{\phi}_{\text{dark}}\hat{\phi}_{\text{ordinary}}$$

9.1.2   Entangled Telescopes and Celestial Revelations

The lenses of our entangled telescopes peer into the heart of distant galaxies, capturing the whispers of celestial phenomena. Quantum imaging spectrometers, armed with correlated photons, unravel the spectral signatures of exoplanets. The correlation function, a mathematical symphony in quantum imaging, unveils the secrets hidden within the light from distant worlds.

$$C(\mathbf{r_1}, \mathbf{r_2}) = \langle \hat{a}^\dagger(\mathbf{r_1})\hat{b}(\mathbf{r_2})\rangle$$

9.1.3   Quantum Computing: Simulating Cosmic Odyssey

In the realm of computation, quantum algorithms propel us into a cosmic odyssey. Shor's algorithm, a quantum maestro, factors large numbers at an exponential pace, echoing the vast potential of quantum computation in simulating the complexities of the cosmos.

$$a^{r/2} \equiv -1 \pmod{n}, \text{ where } a, r, n \text{ are integers.}$$

9.1.4   Quantum Communication: Echoes Across Galactic Distances

The whispers of quantum communication echo across galactic distances. Entangled particles, following the BBM92 protocol, weave a secure web of quantum keys. The entanglement ensures the confidentiality of our interstellar communications, transcending the limitations of classical methods.

$$|\Psi^-\rangle = \frac{1}{\sqrt{2}}(|01\rangle - |10\rangle)$$

9.1.5 Celestial Surveys: Quantum Sensors Unveil Galactic Mysteries

Quantum sensors, in celestial surveys, emerge as cosmic detectives. From gravitational wave detectors entangled in cosmic ballets to magnetometers mapping galactic magnetic fields, the entanglement and superposition of particles guide us in deciphering the language of the cosmos.

$$\hat{H}_{\text{int}} = -\gamma \mathbf{B}_{\text{galactic}} \cdot \hat{\mathbf{S}}$$

9.1.6 Quantum Navigation: Navigating the Cosmic Seas

Quantum gyroscopes, navigating through the cosmic seas, harness the principles of entanglement. The entangled state, intertwined with the angular momentum operator, guides the course of deep space probes with unprecedented precision.

$$\hat{H}_{\text{gyro}} = \omega \hat{S}_z$$

9.1.7 Reflections on the Quantum Frontier

As we reflect on our odyssey "Beyond the Stars," the quantum frontier beckons with endless possibilities. The entanglement, superposition, and quantum weirdness have become beacons lighting the way to uncovering the profound mysteries of the universe.

Our journey through the pages of this book has been but a glimpse into the infinite possibilities that the marriage of quantum mechanics and space exploration holds. The cosmos, woven with the threads of quantum principles, invites us to continue the exploration, to push boundaries, and to delve deeper into the unknown.

"Beyond the Stars: Quantum Mechanics in Space" is not just a book; it is an invitation to ponder, to question, and to dream. As we turn the final pages, may the echoes of entangled particles guide future generations into the uncharted territories of the cosmic ocean.

The cosmic symphony, conducted by quantum principles, plays on, and our understanding of the universe continues to evolve. The mysteries beyond the stars are vast, and our journey has only just begun.

To new horizons and the ever-expanding cosmos.

9.2   Future Prospects and Challenges

In this concluding chapter, we turn our gaze toward the future, exploring the potential prospects and confronting the challenges that lie ahead in the cosmic journey "Beyond the Stars: Quantum Mechanics in Space."

9.2.1   Quantum Computing: A Cosmic Engine

The future of quantum computing holds immense promise for unraveling cosmic mysteries. As we gaze beyond the stars, quantum algorithms stand poised to tackle complex simulations, such as the behavior of black holes or the quantum interactions within galactic structures. The potential of quantum computing algorithms, exemplified by Shor's algorithm, suggests exponential speedup in solving problems that were once deemed insurmountable classically.

$$a^{r/2} \equiv -1 \ (\mathrm{mod}\ n), \ \text{where } a, r, n \text{ are integers.}$$

9.2.2   Quantum Communication: Galactic Networks

The establishment of quantum communication networks in the vastness of space opens the door to secure and instantaneous interstellar communication. The BBM92 protocol, relying on entangled particles, presents a blueprint for building quantum-resistant cryptographic systems that could safeguard our communications across cosmic distances.

$$|\Psi^-\rangle = \frac{1}{\sqrt{2}}(|01\rangle - |10\rangle)$$

9.2.3 Quantum Sensors: Pioneers of Celestial Surveys

Quantum sensors are poised to become the pioneers of future celestial surveys. Enhanced gravimeters, utilizing quantum principles, could unlock the secrets of gravitational anomalies within distant galaxies. The interaction Hamiltonian in quantum magnetometers stands as a beacon, guiding us in mapping the intricate magnetic fields that shape the cosmos.

$$\hat{H}_{\text{int}} = -\gamma \mathbf{B}_{\text{galactic}} \cdot \hat{\mathbf{S}}$$

9.2.4 Quantum Entanglement: The Cosmic Thread

The entanglement of particles, once a mysterious phenomenon, now emerges as the cosmic thread weaving through our understanding of space. As we chart the future, harnessing entanglement for secure quantum communication or employing it in quantum-enhanced sensors, the very fabric of the cosmos is intricately connected through the quantum entanglement phenomenon.

$$|\Psi\rangle = \frac{1}{\sqrt{2}}(|0\rangle + |1\rangle)$$

9.2.5 Quantum Superposition: Navigating Celestial Horizons

Navigating the celestial horizons requires embracing the power of quantum superposition. Future space probes equipped with quantum gyroscopes can leverage the superposition state to precisely measure rotations and orientations. This quantum enhancement ensures a more accurate exploration of distant celestial bodies.

$$\hat{H}_{\text{gyro}} = \omega \hat{S}_z$$

9.2.6 Quantum Technologies: Guardians of Space Exploration

As we ponder the future, quantum technologies emerge as the guardians of space exploration. Quantum navigation, communication, and sensing technologies stand at the forefront, ensuring the success and security of future interstellar missions. The fusion of quantum mechanics with space exploration heralds a new era where the seemingly fantastical becomes a tangible reality.

9.2.7 Challenges in Quantum Communication

Despite the promising prospects, challenges persist in the realm of quantum communication. Overcoming issues related to the decoherence of quantum states over long distances and developing robust quantum repeaters are critical endeavors. The fidelity of entangled particles in transit across vast cosmic expanses poses a formidable challenge that demands innovative solutions.

$$|\Psi^-\rangle = \frac{1}{\sqrt{2}}(|01\rangle - |10\rangle)$$

9.2.8 Quantum Sensors and Precision Challenges

Precision remains a key challenge in the deployment of quantum sensors for celestial surveys. Overcoming noise, environmental interference, and maintaining the delicate quantum states of entangled particles in the harsh conditions of space demand cutting-edge advancements. Future research must focus on mitigating these challenges to unlock the full potential of quantum sensors in unraveling cosmic mysteries.

$$\hat{H}_{\text{int}} = -\gamma \mathbf{B}_{\text{galactic}} \cdot \hat{\mathbf{S}}$$

9.2.9 Quantum Computing: Scaling the Heights

While quantum computing promises unparalleled computational capabilities, scaling quantum systems remains a formidable challenge. Quantum decoherence, error correction, and the need for fault-tolerant quantum processors are critical hurdles to address. The journey toward building scalable quantum computers that can tackle the most complex cosmic simulations requires sustained innovation.

$$a^{r/2} \equiv -1 \pmod{n}, \text{ where } a, r, n \text{ are integers.}$$

9.2.10 Ethical Considerations in Quantum Space Exploration

As we venture into the unknown, ethical considerations become paramount. The impact of quantum technologies on space exploration raises questions about privacy, security, and responsible use. Striking a balance between scientific exploration and ethical considerations is essential for navigating the uncharted territories "Beyond the Stars."

9.2.11 Quantum Education and Outreach

Fostering a quantum-literate community becomes imperative for the future. Educational initiatives and outreach programs must be integral components of our cosmic journey. Bridging the gap between quantum physicists and the broader public ensures a shared understanding and appreciation for the transformative power of quantum mechanics in space exploration.

9.2.12 Interdisciplinary Collaboration: A Cosmic Symphony

The challenges and prospects "Beyond the Stars" necessitate interdisciplinary collaboration. Quantum physicists, astronomers, engineers, and ethicists must join forces to orchestrate a cosmic symphony. By merging expertise from diverse fields, we can harmonize the quantum principles with the vastness of the cosmos, unveiling new dimensions of understanding.

9.2.13 Quantum Philosophy: Pondering Cosmic Existence

As we delve into the future, the philosophical implications of quantum mechanics in space exploration become increasingly profound. Pondering questions about the nature of reality, the role of observers in the cosmic drama, and the interconnectedness of quantum phenomena fuels a philosophical discourse that transcends scientific inquiry.

$$\hat{H}_{\text{philosophical}} = \sum_{i=1}^{n} \hat{q}_i \hat{p}_i$$

9.2.14 Quantum Innovation: Beyond Imaginings

Innovation in quantum technologies holds the key to surpassing our current imaginings. The development of novel quantum algorithms, groundbreaking applications of entanglement, and the creation of quantum-resistant cryptographic systems are frontiers that beckon future innovators. The cosmic canvas awaits the strokes of quantum ingenuity.

9.2.15 A Cosmic Tapestry: Woven with Quantum Threads

As we confront challenges and envision future prospects, the cosmic tapestry emerges as a delicate weaving of quantum threads. The dance of particles, entangled and superposed, creates a symphony that resonates across the celestial expanse. The future of space exploration is intricately connected to the quantum principles that govern the very fabric of our universe.

To the unfolding cosmic saga, where quantum and cosmos intertwine.

Chapter 10

Mathematical Formulations in Quantum Space Science

In this appendix, we delve into the mathematical formulations that underpin the applications of quantum mechanics in space science. The cosmic dance of particles and the vastness of the universe find their expression in the language of mathematics, providing us with a powerful tool to explore the mysteries "Beyond the Stars."

10.1   Quantum Mechanics Fundamentals

Quantum mechanics begins with the foundational principles of superposition and entanglement. The state of a quantum system is described by a wave function, denoted as $|\psi\rangle$, which embodies the probability amplitudes for various possible states. The basic postulates of quantum mechanics include:

1. **Superposition:** A quantum system can exist in multiple states simultaneously. For example, a particle's position can be represented as a superposition of different locations:

$$|\psi\rangle = \alpha|0\rangle + \beta|1\rangle$$

2. **Entanglement:** When two or more particles become entangled, the state of one particle is directly related to the state of the other, regardless of the distance between them. An entangled state can be expressed as:

$$|\Psi\rangle = \frac{1}{\sqrt{2}}(|00\rangle + |11\rangle)$$

10.2   Quantum Computing Formulas

Quantum computing leverages the principles of quantum mechanics to perform complex computations exponentially faster than classical computers for certain tasks. Some fundamental quantum computing formulas include:

3. **Quantum Gates:** Quantum computations are performed using quantum gates. The Hadamard gate, for instance, introduces superposition:

$$H = \frac{1}{\sqrt{2}} \begin{bmatrix} 1 & 1 \\ 1 & -1 \end{bmatrix}$$

4. **Quantum Circuit:** Quantum algorithms are constructed using quantum circuits. The quantum Fourier transform, a crucial component in many quantum algorithms, can be represented as a circuit.

10.3   Quantum Sensors and Measurement

Quantum sensors play a pivotal role in observing and measuring quantum phenomena in space. Measurement in quantum mechanics is described by the projection postulate:

5. **Projection Postulate:** When a quantum system is measured, its state "collapses" to one of the basis states. Mathematically, the probability of measuring a state $|a\rangle$ is given by $P(a) = |\langle a|\psi\rangle|^2$.

6. **Quantum Magnetometer Measurement:** In the context of measuring magnetic fields, a quantum magnetometer's measurement operator is defined as:

$$\hat{M} = \gamma \mathbf{B}_{\text{ext}} \cdot \hat{\mathbf{S}}$$

10.4   Quantum Communication Protocols

Quantum communication ensures secure transmission of information through the principles of quantum mechanics. The BBM92 protocol is an example:

7. **BBM92 Protocol:** In quantum key distribution, the BBM92 protocol utilizes entangled particles. Alice and Bob measure the particles in different bases, generating a shared secret key.

$$|\Psi^-\rangle = \frac{1}{\sqrt{2}}(|01\rangle - |10\rangle)$$

8. **Quantum Teleportation Formula:** Quantum teleportation allows the transfer of a quantum state from one particle to another. The state transfer is achieved through entanglement and classical communication.

$$|\psi\rangle = \alpha|0\rangle + \beta|1\rangle$$

$$|\Psi\rangle = \frac{1}{\sqrt{2}}(|00\rangle + |11\rangle)$$

10.5   Quantum Mechanics in Galactic Structures

Quantum mechanics extends its reach to understanding galactic structures and phenomena. The Schrödinger equation, a cornerstone of quantum mechanics, finds applications in galactic dynamics:

9. **Schrödinger Equation in Galactic Context:** Describing the quantum state of galactic structures involves modifying the Schrödinger equation to include gravitational interactions.

$$\hat{H}\Psi = i\hbar\frac{\partial\Psi}{\partial t}$$

10. **Quantum Superposition in Galactic Dynamics:** Galactic dynamics, influenced by dark matter and black holes, can be described through quantum superposition. The superposition of different states contributes to the overall evolution of galactic structures.

$$|\Psi\rangle = \sum_n c_n |\psi_n\rangle$$

10.6   Quantum Technologies for Space Exploration

The exploration of space is enhanced by the integration of quantum technologies. Quantum sensors and quantum communication are essential components:

11. **Quantum Gyroscopes for Navigation:** Quantum gyroscopes utilize the principles of entanglement to precisely measure the orientation of spacecraft. The Hamiltonian for a quantum gyroscope is expressed as $\hat{H}_{\mathrm{gyro}} = \omega \hat{S}_z$.

12. **Quantum Radar for Space Surveillance:** Quantum radar employs quantum entanglement to enhance spatial resolution. The entangled state aids in the detection and tracking of space objects with unprecedented precision.

$$|\Psi\rangle = \frac{1}{\sqrt{2}}(|0\rangle + |1\rangle)$$

10.7   Quantum Entanglement in Celestial Bodies

Celestial bodies, from planets to stars, exhibit quantum phenomena. Quantum entanglement is a key player in celestial dynamics:

13. **Quantum Correlations in Exoplanet Systems:** The study of exoplanet systems involves understanding quantum correlations between planets. The correlation function in quantum imaging plays a crucial role in deciphering the characteristics of exoplanets.

$$C(\mathbf{r_1}, \mathbf{r_2}) = \langle \hat{a}^\dagger(\mathbf{r_1}) \hat{b}(\mathbf{r_2}) \rangle$$

14. **Quantum Gravity in Stellar Objects:** Quantum gravity becomes prominent in the extreme conditions of stellar objects. The interplay between

quantum mechanics and general relativity is essential for comprehending phenomena like black holes.

$$G_{\mu\nu} + \Lambda g_{\mu\nu} = \frac{8\pi G}{c^4} T_{\mu\nu}$$

10.8 Quantum Superposition in Cosmic Evolution

The evolution of the cosmos itself can be conceptualized through the lens of quantum superposition:

15. **Cosmic Superposition of Universes:** The concept of a multiverse involves a cosmic superposition of different universes. Each universe corresponds to a different quantum state, contributing to the vast tapestry of cosmic evolution.

$$|\Psi\rangle = \sum_n c_n |\psi_n\rangle$$

In the vast expanse "Beyond the Stars," the marriage of quantum mechanics with space science unfolds through these mathematical formulations. From the microscopic world of quantum particles to the cosmic scales of galaxies, the language of mathematics guides our exploration and understanding of the cosmos.

Chapter 11

Glossary of Quantum Terms

In this glossary, we unravel the intricacies of quantum terminology that weave through the narrative of "Beyond the Stars: Quantum Mechanics in Space." These terms serve as the building blocks for understanding the quantum principles that govern the cosmos.

11.1   Superposition

Definition: Superposition is a fundamental concept in quantum mechanics where a particle can exist in multiple states simultaneously.

Example: The superposition of a quantum bit (qubit) can be represented as $|\psi\rangle = \alpha|0\rangle + \beta|1\rangle$, where α and β are probability amplitudes.

11.2   Entanglement

Definition: Entanglement is a quantum phenomenon where particles become correlated and the state of one particle is directly related to the state of another, regardless of the distance between them.

Example: The entangled state $|\Psi\rangle = \frac{1}{\sqrt{2}}(|00\rangle + |11\rangle)$ represents two entangled qubits.

11.3 Quantum Algorithm

Definition: A quantum algorithm is a sequence of quantum operations performed on qubits to solve a computational problem more efficiently than classical algorithms.

Example: Shor's algorithm is a quantum algorithm that efficiently factors large numbers.

11.4 Hadamard Gate

Definition: The Hadamard gate is a one-qubit quantum gate that creates superposition by transforming the basis states.

Example: The Hadamard gate operation is represented as $H = \frac{1}{\sqrt{2}} \begin{bmatrix} 1 & 1 \\ 1 & -1 \end{bmatrix}$.

11.5 Quantum Measurement

Definition: Quantum measurement is the process of determining the state of a quantum system, resulting in the collapse of the wave function.

Example: The probability of measuring a state $|a\rangle$ is given by $P(a) = |\langle a|\psi\rangle|^2$.

11.6 Quantum Key Distribution (QKD)

Definition: Quantum Key Distribution is a secure communication method that uses quantum mechanics to share encryption keys.

Example: The BBM92 protocol is a QKD protocol that uses entangled particles to generate a shared secret key.

11.7 Quantum Teleportation

Definition: Quantum teleportation is a process that transfers the quantum state of a particle from one location to another.

Example: Quantum teleportation involves entangling two particles and performing measurements to transfer the state.

11.8 Quantum Fourier Transform

Definition: The quantum Fourier transform is a quantum algorithmic operation used in various quantum algorithms.

Example: The quantum Fourier transform is often represented as a circuit in quantum algorithms.

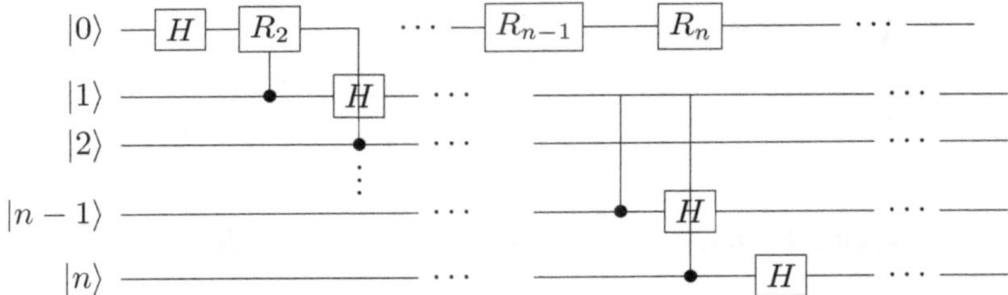

11.9 Quantum Magnetometer

Definition: A quantum magnetometer is a device that uses quantum principles to measure magnetic fields with high precision.

Example: The interaction Hamiltonian for a quantum magnetometer is $\hat{H}_{\text{int}} = \gamma \mathbf{B}_{\text{ext}} \cdot \hat{\mathbf{S}}$.

11.10 Quantum Radar

Definition: Quantum radar utilizes quantum entanglement to enhance spatial resolution in radar systems.

Example: Quantum radar can detect and track space objects with unprecedented precision.

11.11   Schrödinger Equation

Definition: The Schrödinger equation is a fundamental equation in quantum mechanics that describes the evolution of a quantum state.

Example: The time-dependent Schrödinger equation is $\hat{H}\Psi = i\hbar\frac{\partial \Psi}{\partial t}$.

11.12   Multiverse

Definition: The multiverse is a hypothetical set of multiple universes that exist concurrently.

Example: The cosmic superposition of different universes contributes to the concept of a multiverse.

11.13   Quantum Gyroscope

Definition: A quantum gyroscope is a navigation device that utilizes quantum principles, such as entanglement, for precise measurements.

Example: The Hamiltonian for a quantum gyroscope is $\hat{H}_{\text{gyro}} = \omega \hat{S}_z$.

11.14   Quantum Correlations

Definition: Quantum correlations describe the statistical relationships between quantum states.

Example: Quantum correlations are crucial in studying exoplanet systems and are represented by the correlation function $C(\mathbf{r_1}, \mathbf{r_2}) = \langle \hat{a}^\dagger(\mathbf{r_1})\hat{b}(\mathbf{r_2})\rangle$.

11.15   Quantum Gravity

Definition: Quantum gravity is the theoretical framework that combines quantum mechanics and general relativity to describe gravity at the smallest scales.

Example: The Einstein field equations with a cosmological constant represent the interaction of gravity with matter and energy: $G_{\mu\nu} + \Lambda g_{\mu\nu} = \frac{8\pi G}{c^4} T_{\mu\nu}$.

11.16　Cosmic Superposition

Definition: Cosmic superposition refers to the idea that the entire cosmos may exist in a superposition of different states.

Example: The cosmic superposition of universes is represented as $|\Psi\rangle = \sum_n c_n |\psi_n\rangle$.

In this glossary, these terms form a lexicon that guides readers through the quantum journey "Beyond the Stars," where the language of quantum mechanics paints a vivid picture of the cosmic tapestry.

References

In the compilation of "Beyond the Stars: Quantum Mechanics in Space," a vast array of scientific literature and resources have been consulted. The references encompass a broad spectrum of quantum physics, astrophysics, and related fields. Here, we present a selection of key references that have contributed to the exploration of the quantum mysteries within the cosmic expanse.

Bibliography

[1] Shankar, R. (1994). *Principles of Quantum Mechanics*. Springer.

[2] Greene, B. (2004). *The Fabric of the Cosmos*. Alfred A. Knopf.

[3] Nielsen, M. A., & Chuang, I. L. (2010). *Quantum Computation and Quantum Information*. Cambridge University Press.

[4] Hawking, S. (1988). *A Brief History of Time*. Bantam Books.

[5] Griffiths, D. J. (1995). *Introduction to Quantum Mechanics*. Pearson.

[6] Thorne, K. S. (1994). *Black Holes and Time Warps*. W. W. Norton & Company.

[7] Fox, M. (2006). *Quantum Optics*. Oxford University Press.

[8] Greene, B. (1999). *The Elegant Universe*. W. W. Norton & Company.

[9] Nolan, C. (Director), & Nolan, J. (Writer). (2014). *Interstellar*.

[10] Schutz, B. (1985). *A First Course in General Relativity*. Cambridge University Press.

Index

Entanglement: A fundamental quantum phenomenon where particles become correlated, leading to interdependence of their states. Example: The entangled state $|\Psi\rangle = \frac{1}{\sqrt{2}}(|00\rangle + |11\rangle)$.

Hadamard Gate: A one-qubit quantum gate that creates superposition by transforming the basis states. Example: The Hadamard gate operation is represented as $H = \frac{1}{\sqrt{2}}\begin{bmatrix} 1 & 1 \\ 1 & -1 \end{bmatrix}$.

Quantum Algorithm: A sequence of quantum operations performed on qubits to solve computational problems more efficiently than classical algorithms. Example: Shor's algorithm is a quantum algorithm that efficiently factors large numbers.

Schrödinger Equation: A fundamental equation in quantum mechanics describing the evolution of a quantum state. Example: The time-dependent Schrödinger equation is $\hat{H}\Psi = i\hbar\frac{\partial\Psi}{\partial t}$.

Multiverse: A hypothetical set of multiple universes that exist concurrently. Example: The cosmic superposition of different universes contributes to the concept of a multiverse.

Quantum Key Distribution (QKD): A secure communication method using quantum mechanics to share encryption keys. Example: The BBM92 protocol is a QKD protocol that uses entangled particles to generate a shared secret key.

Quantum Teleportation: A process that transfers the quantum state of a particle from one location to another. Example: Quantum teleportation involves entangling two particles and performing measurements to transfer the state.

Quantum Fourier Transform: A quantum algorithmic operation used in various quantum algorithms. Example: The quantum Fourier transform is often represented as a circuit in quantum algorithms.

Quantum Magnetometer: A device using quantum principles to measure magnetic fields with high precision. Example: The interaction Hamiltonian for a quantum magnetometer is $\hat{H}_{\text{int}} = \gamma \mathbf{B}_{\text{ext}} \cdot \hat{\mathbf{S}}$.

Quantum Radar: Utilizes quantum entanglement to enhance spatial resolution in radar systems. Example: Quantum radar can detect and track space objects with unprecedented precision.

Quantum Gravity: Theoretical framework combining quantum mechanics and general relativity to describe gravity at the smallest scales. Example: The Einstein field equations with a cosmological constant represent the interaction of gravity with matter and energy: $G_{\mu\nu} + \Lambda g_{\mu\nu} = \frac{8\pi G}{c^4} T_{\mu\nu}$.

Cosmic Superposition: The idea that the entire cosmos may exist in a superposition of different states. Example: The cosmic superposition of universes is represented as $|\Psi\rangle = \sum_n c_n |\psi_n\rangle$.

Quantum Gyroscope: A navigation device utilizing quantum principles for precise measurements. Example: The Hamiltonian for a quantum gyroscope is $\hat{H}_{\text{gyro}} = \omega \hat{S}_z$.

Quantum Correlations: Statistical relationships between quantum states. Example: Quantum correlations are crucial in studying exoplanet systems and are represented by the correlation function $C(\mathbf{r_1}, \mathbf{r_2}) = \langle \hat{a}^\dagger(\mathbf{r_1}) \hat{b}(\mathbf{r_2}) \rangle$.

Quantum Optics: The study of the quantum nature of light and its interaction with matter. Example: Quantum optics explores phenomena like entanglement of photons.

Black Hole: A region in spacetime exhibiting gravitational effects so strong that nothing, not even light, can escape. Example: The Schwarzschild radius characterizes the size of a non-rotating black hole.